Francis E. Nipher

Theory of Magnetic Measurements

With an Appendix on the Method of Least Squares

Francis E. Nipher

Theory of Magnetic Measurements
With an Appendix on the Method of Least Squares

ISBN/EAN: 9783743383814

Manufactured in Europe, USA, Canada, Australia, Japa

Cover: Foto ©berggeist007 / pixelio.de

Manufactured and distributed by brebook publishing software
(www.brebook.com)

Francis E. Nipher

Theory of Magnetic Measurements

THEORY

OF

MAGNETIC MEASUREMENTS,

WITH AN APPENDIX ON THE

METHOD OF LEAST SQUARES.

BY

FRANCIS E. NIPHER, A.M.,

Professor of Physics in Washington University, President of the St. Louis Academy of Science.

NEW YORK

D. VAN NOSTRAND, PUBLISHER,
23 MURRAY AND 27 WARREN STREETS.
1886.

H. J. HEWITT, PRINTER AND ELECTROTYPER,
27 ROSE STREET, N. Y.

PREFACE.

During the last four or five years the writer has frequently been requested to furnish information relating to the practical details of a magnetic survey. The need of a brief hand-book to supplement the instructions of the Coast and Geodetic Survey was felt by the writer while preparing for a similar survey of Missouri, and this little volume is offered to the public in the hope that it may be of service to others contemplating similar work.

At the same time the great and growing importance of electrical and magnetic measurements will perhaps commend the volume to a wider circle of readers.

The discussion of the method of Least Squares is an extension of an article in Weisbach's "Mechanics."

F. E. N.

St. Louis, May 24, 1886.

CONTENTS.

INTRODUCTION.

A FORCE is completely known when its direction and intensity have been determined.

The direction of the lines of force in the earth's gravitation field is indicated by the plumb-line. The plumb-bob tends to move downwards along the line of force. The direction of the lines of force of the earth's magnetic field would be indicated by the direction taken by the magnetic axis of a magnetic needle suspended at its centre of gravity, so that it could move freely in any direction. When thus placed the opposite ends of the magnet-needle tend to move in opposite directions along the line of force. The magnetic axis of a needle is a line passing through its poles, as will be explained more fully later.

The position of the lines of force is referred to two planes—viz., the horizontal plane and the plane of the geographical meridian. The position of the lines of force is really determined by means of two magnet-needles, one of which is free to move in a horizontal plane, and the other in a vertical plane, since it is impossible mechanically to combine these motions in a single needle of sufficient delicacy.

The first needle determines, at the point of observation, the angle between the plane of the geographical meridian and the vertical plane containing the line of force at the point. This angle is called the declination.

The second needle determines the angle between the

line of force and a horizontal plane.   This angle is called the inclination or dip.   It remains to explain the method used in determining the intensity of force.

In earlier times the weight of a unit mass was taken as the unit of force.   The weight of a pound is not the same for different places on the earth, and hence this unit can only be used for rough work or for local determinations.

According to Newton's law for attraction, the force with which either of two masses m and M attract the other, the distance between them being d, is

$$w = K' \frac{Mm}{d^2}. \qquad (1)$$

This equation is verified by the motions of the planets and of falling bodies.   If M represents the mass of the earth in pounds or grammes, m being the mass of a body at the surface, d being the radius of the earth, in feet or centimetres, then w represents the force (measured in units which are not yet supposed to be fixed) by which a spring separating the two bodies would be compressed. This force is usually called the weight of the mass m. It is equally the weight of the earth.   The weight of the earth is always equal to that of the body weighed, and is therefore an indeterminate quantity. K' represents the force with which a unit mass would attract another unit mass, the distance between them being unity.

By experiments with Attwood's machine it is shown that if a force acts upon m units of mass, imparting an acceleration a, the force is represented by the expression,

$$F = K\,ma, \qquad (2)$$

where K is the force which would impart a unit acceleration to a unit mass, F being determined in any units.

If the body m is free to move under the attraction of the earth, it receives an acceleration g. Hence the attraction of the earth for the m units of mass is

$$w = K \, mg. \qquad (3)$$

The earth, having a mass M, likewise moves with an acceleration a' towards the body m, the force acting on the earth being therefore

$$K' \, Ma' = w = K' \, mg,$$

so that

$$a' = \frac{m}{M} g.$$

The value $\dfrac{m}{M}$ is so small that a' is inappreciable.

In equation (1) it is evident that the value of K' might be taken as the unit of force. The value K in (2) and (3) would not then be unity. This is not done, however, but the unit of force is so chosen that K in (2) and (3) is unity. The unit force is then that force which can impart a unit acceleration to a unit mass (gramme or pound).

In centimetre-gramme-second (abbreviated C. G. S.) units this unit of force is called the dyne. Equation (3) then becomes

$$w = m . g, \qquad (4)$$

where w is the weight of m units of mass expressed in the above-chosen unit. It also follows that

$$\frac{w}{m} = g.$$

The left side of this expression is the weight of a unit mass. In English units the weight of a pound at London is 32.1912; at any other place the weight of a pound is g units of force, where g is the acceleration of a falling body at the place.   In C. G. S. units the weight of a gramme at Paris is 980.94 dynes, and at any point it is g dynes.   The weight of $\frac{1}{g}$ units of mass is therefore the unit force.

The weight of a gramme at any point in the earth's field gives a measure of the force which acts upon the gramme at that point, tending to cause motion along the lines of force ; in other words, the value of g at any point is a measure of the strength of the earth's gravitation field at that point.

Adopting the above unit of force, the value of K' in (1) may be calculated as follows in C. G. S. units : Consider the case of a gramme at the surface of the earth, M representing the mass of the earth in grammes, and d its mean radius in centimetres.   Then in (1) $m = 1$; $M = 6.14 \times 10^{27}$; $d = 6.37 \times 10^{8}$; $w = 981$.   Hence

$$K = \frac{981 \times (6.37 \times 10^{8})^{2}}{6.14 \times 10^{27}} = \frac{1}{1.543 \times 10^{7}}.$$

From this the mass which must be placed at any point in order to attract an equal mass with a force of one dyne can be computed, since

$$1 = K' \frac{mm}{1^{2}}$$

$$m = \sqrt{\frac{1}{K'}} = 3928 \text{ grammes.}$$

Hence 3928 grammes would attract an equal mass at a distance of one centimetre, with a force which would impart to one gramme an acceleration of a centimetre per second, or to 3928 grammes an acceleration of $\frac{1}{3928}$ cm. per sec.   The 3928 grammes is called the as-

tronomical unit of mass. If this is taken as the unit
mass the astronomical equation of attraction becomes

$$w = \frac{mM}{d^2},\qquad (5)$$

w being given in the ordinary unit of force—the dyne,
or some equivalent unit.

The unit magnetic pole or the electro-static unit
quantity of electricity is defined in accordance with this
equation. The unit pole is that pole which will act upon
an equal pole at a unit distance with a unit force.

DECLINATION.

The magnetic needle commonly used for precise determinations is of the collimator form, consisting of a
small, cylindrical shell of steel, one end of which is
closed by a lens, the principal focus of which is on a
scale, etched or photographed on a glass plate, which
closes the other end. This scale should be divided de-
cimally into about 100 parts, numbered continuously
from one extremity to the other. The angle subtended
at the middle of the magnet by one scale division should
be between one and three minutes. This angle is called
the scale value of the magnet.

The magnet hangs in a stirrup, supported on a long
fibre of raw silk, its position in the stirrup being fixed
by small brass guide-rings around the magnet. The
magnet is enclosed in a box, from the top of which a
glass tube extends upward, its top terminating in a
graduated torsion-head to which the suspension-fibre
is attached. The ends of the box are provided with

windows, one admitting light from a mirror upon the scale, and the other for telescopic observation. It is usually better to use well-seasoned wood in the construction of the box, and to avoid the use of metal in all parts nearest to the magnet, as it is almost impossible to obtain brass free from iron. It is necessary to examine all brass screws or fittings in a declinometer, if they are near the magnet, and, if found to be magnetic, they should be replaced by others, or proper correction made for their effect on the position of the needle and on the intensity of the field. This correction, although really a function of the strength of the field, may be taken as constant.

In some forms of instrument the observing telescope is connected with the magnet-box, and mounted with it on the same azimuth circle, the centre of which is below the point of suspension of the magnet and in the same vertical. In others it consists of a transit, or altazimuth instrument, mounted on the fixed support which carries the magnet-box. In the former case a change in the pointing of the telescope introduces a torsion in the fibre. The latter instrument, which is due to Gauss, is preferable for field-work.

DETERMINATION OF SCALE VALUE OF THE MAGNET.

Let the telescope be focussed upon the centre of the magnet scale, in which case we may assume, in order to fix our ideas, that the optical axes of the telescope and the collimator magnet coincide. If the magnet be turned on its suspension-fibre through any angle, and the telescope be turned on its vertical axis through the same angle and in the same direction, the optical axes of the magnet and telescope will again be parallel. If the two instruments turned about a common vertical axis the optical axes will also be coincident; but if they turn around parallel axes the optical axes will be parallel and not coincident in the second position. In both cases, however, the scale-reading in the first and second positions will be the same. This preservation of an unchanged scale-reading in the case mentioned is also true if the lines of collimation of the magnet and telescope are not parallel. The value of one division of the scale may therefore be determined by pointing the telescope successively on the principal divisions of the scale, taking the readings of the azimuth circle for each pointing. If the magnet-box moves with the telescope the magnet must hang on its suspension-fibre, and the readings must be corrected for torsion in a manner to be hereafter explained. If the magnet-box does not move the magnet may be fastened in its normal position during the operation. If the number of pointings is odd, the circle-readings corresponding to divisions equidistant from the middle of the scale are reduced to the mean division by finding their means, as is shown in the third column of the table below. The mean reading of the middle division is here 338° 53'.4.

DETERMINATION OF SCALE VALUE OF MAGNET C_t.

Scale.	Azimuth circle. Mean of Verniers.	Reading of middle division.	DIFFERENCE FROM MEAN.	
			Circle.	Scale.
160	335° 44′.7	....	188.7	80
150	336 09.0	....	164.4	70
140	336 34.0	....	139.4	60
130	336 57.5	....	115.9	50
120	337 19.5	....	93.9	40
110	337 43.0	....	70.4	30
100	338 06.0	....	47.4	20
90	338 29.5	....	23.9	10
80	338 52.5	338° 52′.5	00.9	00
70	339 16.5	53.0	23.1	10
60	339 39.5	52.7	46.1	20
50	340 05.0	54.0	71.6	30
40	340 27.5	53.5	94.1	40
30	340 52.7	55.1	119.3	50
20	341 15.6	54.8	142.2	60
10	341 37.5	53.3	164.1	70
0	342 00.0	52.3	186.6	80
		338° 53′.4	Sum 1692.0	Sum 720

The scale value is therefore $\frac{1692}{720} = 2'.35.$

DETERMINATION OF THE MAGNETIC AXIS.

The magnetic axis is a straight line joining the poles of the magnet. If the magnet is freely suspended the axis lies in the line of force. The magnetic axis is determined by taking scale-readings with the scale alternately erect and inverted. If the line of collimation of the telescope should happen to coincide with the magnetic axis of the magnet, it would then be pointed in the plane of the magnetic meridian, and the reading of

the erect and the inverted scale would be identical.   The
division-lines of the scale should always be accurately
vertical when read.   The position of the magnetic axis
varies continually by the jarring incident to travel.   A
freshly-magnetized magnet is especially unstable, chang-
ing even with variations of temperature.   Declination
magnets should be carried south end up, and should be
kept away from other magnets.   The axis should be determined at each station, or as often as experience
shows to be necessary.   This will, of course, depend
upon the age of the magnet, the hardness of the steel,
and the kind of treatment the magnet receives.   The
custom which some county surveyors have of remagnet-
izing their compass-needles whenever they are dissatis-
fied with their instrument is not a wise custom.

MAGNETIC AXIS OF C_6.

Magnet.	SCALE.		Mean.	Alternate means, 1 and 3, 2 and 4, etc.	Axis reads.
	Left.	Right.			
E	74.6	75.1	74.8	...	...
I	79.9	82.3	81.1	75.00	78.05
E	62.8	87.7	75.2	81.00	78.10
I	74.9	87.0	80.9	75.25	78.07
E	64.6	86.1	75.3	80.80	78.05
I	70.4	91.0	80.7	75.50	78.10
E	65.0	86.5	75.7	...	
					78.07

When the telescope is pointed on the division 78.07
of the scale its line of collimation is in the plane of
the magnetic meridian.   The left and right scale-readings in columns 2 and 3 are the extreme readings of the
scale during an oscillation, it being assumed that the
amplitude does not diminish.   This is sufficiently precise for heavy needles or small amplitudes.

The determination of magnetic declination involves a determination of the direction of a true north and south line, as shown by the azimuth circle, and the mean position of the magnetic axis of the needle for the day, or for a series of days, as read on the same circle. The daily swing of the needle in summer is, on the average, about 15′, the north end of the needle being at its greatest eastern elongation at about 7.15 to 7.30 o'clock A.M. on normal summer days, and at its western elongation at 1.15 to 2 o'clock P.M. These hours vary somewhat with the season of the year and for different parts of the country. The mean position of the needle, as deduced from hourly observations throughout the day, is, on the average, within half a minute of the mean of eastern and western elongations. The mean for successive days frequently varies by five minutes, even in times of minimum magnetic disturbance.

This mean position might be obtained by pointing on the axis reading of the magnet at the two elongations, and taking the mean of the azimuth circle-readings. It is better to point approximately on the axis at about 6.30 A.M., and, clamping the circle, to take readings of the magnet-scale at intervals of ten or fifteen minutes until after elongation has passed. Leaving the circle unchanged, if possible, a similar set of scale-readings should be taken, so as to include the afternoon elongation. These observations, then, show any abnormal changes in declination. The mean scale-reading of the two elongations is then found ; if it should happen to coincide with the magnetic axis the telescope would then be pointed in the mean magnetic meridian for the day. Should it not thus coincide the circle-reading must then be corrected by the small angle over which the telescope would sweep in turning from the mean scale-reading of the elongations to the magnetic axis. The sign of this correction will depend upon whether the scale is erect or inverted, whether it reads from left to right or the reverse when the scale is erect, and whether the telescope shows an erect or an inverted image. The silk fibre should be examined both before and after each set of observations, in order to detect any torsion that may develop. Changes in atmospheric humidity are likely to develop torsion in a fibre, particularly if it be a new one. The fibre should be no larger than is necessary to sustain the magnet without too frequent breaking. To examine for torsion, the magnet is removed from the stirrup and a brass cylinder of the same weight substituted. The torsion-head should then be turned until this cylinder sets parallel to the magnet-box.

Before the magnetic observations are begun the verniers of the transit should be set to 0° and 180°, and the instrument pointed on some well-defined object near the horizon. The lower clamp being secured, the upper one is released and the magnetic observations begun as explained. The mark should be far enough away so that it will be unnecessary to refocus when the telescope is pointed on the scale. In choosing a mark it is well to remember that objects easily visible in the evening may be invisible in the morning by reason of fogs or changes in shadows. The station should always be described by aid of sketches, and the distance and bearing of corner-stones or other available points of reference recorded. The following table shows the method of recording the observations. Specimen blanks for magnetic observations can be obtained by observers on addressing the Superintendent of the U. S. Coast and Geodetic Survey, Washington, D. C.

1*

MAGNETIC DECLINATION AT JEFFERSON CITY, MO.

In orchard of Phil. E. Chappell. Mark—spire of State House, about one mile distant. Date, Aug. 12, 1879. Instrument, Declinometer No. 3, U. S. C. and G. Survey. Magnet No. 1, scale erect. Scale value, 1'.90. Mark reads, A, 180° 00'.0 ; B, 359° 58'.5 at 6 A.M. Line of detorsion, 15°. Azimuth circle set to A 363° 56'.0 ; B 183° 55'.0. Observer, F. E. N.

Time.	SCALE-READING.		Mean.	Remarks.
	Left.	Right.		
A.M.				
7ʰ 18ᵐ	81ᵈ.1	8ᵈ.05	83.05	Removed torsion weight at 7ʰ 02ᵐ A.M.
7 23	. . .	. . .	83.2	
7 37	83 .0	83 .6	83.3	
7 55	82 .4	84 .1	83.3	
8 17	82 .9	83 .9	83.4	Max. East.
8 30	82 .5	83 .0	82.75	
8 40	. . .	. . .	83.0	
9 25	81 .9	82 .9	82.4	Removed magnet. Line of detorsion unchanged. Replaced magnet.
P.M.				
1ʰ 15ᵐ	78ᵈ.2	79ᵈ.0	78.6	Line of detorsion same. Azimuth circle not changed.
?				
1 25	78 .3	78 .9	78.6	
1 40	78 .2	78 .9	78.55	
1 52	. . .	. . .	78.5	Max. West.
2 07	78 .2	79 .2	78.7	
2 16	. . .	. . .	78.8	
2 30	78 .7	79 .1	78.9	Line of detorsion, 15°.

Mark reads, A 180° 01'.0 ; B 360° 00'.0.

Mean reading E. and W. elongations..		$80^d.9$
Axis of magnet reads................		80 .8
Reduction to axis....................		$+0.1=+0'.2$
Azimuth circle reads................	A	$363°$ $55'.0$
Magnetic south reads................		363 55 .2
Mark reads.........................		$180°$ $00'.0$
Azimuth of mark....................	S	175 28 .1 E
True south reads		355 28 .1
Magnetic declination E. of N...........		$8°$ $27'.1$

In illuminating the magnet-scale the direct solar ray should never, under any circumstances, be allowed to enter the magnet-box.   An illumination from a white cloud or an illuminated sheet of paper is effective.   In field-work, if a tent is not available, the whole instrument should be covered with a soft, heavy cloth, and the tripod should similarly be protected against solar radiation.   No vibrations of the magnet should be allowed, excepting small vibrations about a vertical axis.   Larger vibrations may be checked by the end of the finger ; but it is better to use a camel's-hair brush operated from the outside by means of a lever or spring, which must be so arranged that no air-currents are introduced into the magnet-box.   In field-work the tripod must always be mounted firmly on stakes.   Unless the observations of a survey are made during intervals of magnetic calms, it is necessary to establish a base station, where all observations of declination as well as intensity are observed at the same time as at the field station.

The methods for determining the true meridian are easily accessible, and it is not thought necessary to treat this part of the subject.   For the work of a survey a precision of one minute of arc is sufficient.   Star obser-

vations are the most satisfactory. The method of equal
altitudes requires more time than it is sometimes convenient to devote to it.   The best method of determining
the meridian is by observation of a circumpolar star on
elongation.

Table I., at the end of this volume, gives the time of
occurrence of the elongation of the pole-star, correct
within five minutes, for the years between 1885 and 1895.
The time is local astronomical time.

Table IV., at the close of this volume, gives the azimuth
of Polaris at elongation (counted from the north) for
the years 1885 to 1895, inclusive, and is accurate enough
for all ordinary purposes.

It should be observed, however, that in computing this
table the mean declination of Polaris for the beginning
of the year is necessarily used.   To obtain the azimuth
to the nearest tenth of a minute the apparent declination for the date of observation should be employed, and
the azimuth computed from the following formula :

$$\sin A = \frac{\cos \delta}{\cos \varphi} \qquad (6)$$

where A = azimuth, δ = declination, and φ = latitude.
The apparent place of Polaris is given in the American
Ephemeris for every day in the year.   The difference in
the mean and apparent places may produce a difference
of $0'.5$ in the computed azimuth, and values of the azimuth taken from the table are therefore subject to an
error of that amount.

The precision with which the true meridian must be
determined depends upon the precision with which the
magnetic meridian is determined—or, in other words,
upon the number of days of observation.

In order to illuminate the field of the instrument

where there is no axial illumination, a minute mirror may be mounted in front of the object-glass of the telescope. This mirror may be mounted on a ring clasping the tube. The position of the mirror can be made adjustable by a double-jointed rod attached to the ring. A bull's-eye lantern will then serve to throw light upon the mirror.

It is, of course, always understood that the observing telescope is always kept in adjustment, the levels and line of collimation being carefully examined. For more careful determinations the telescope should be reversed ; but in the field-work of a survey this is usually unnecessary if the instrument is kept in adjustment.

INCLINATION.

The inclination or dip is the angle between the line of force and a horizontal line lying in the magnetic meridian. A needle accurately balanced on a horizontal axis directed in the magnetic prime-vertical, so that the needle moves freely in the plane of the magnetic meridian, will, when magnetized, set with its magnetic axis in the line of force at the point. The position of the needle is determined by means of a graduated circle, having its plane coincident with that of the magnetic meridian. The zero of graduation is a horizontal diameter. In some instruments the ends of the needle point to the scale divisions, which are read by means of magnifying-glasses. In such instruments the graduation is not closer than to ten minutes. In other instruments the observation is made by means of compound microscopes having radially-placed threads, which are set on the marked ends of the needle, the position being determined by means of verniers.

The circle of the instrument is placed in the magnetic meridian by means of a long, horizontal compass-needle set parallel to some mark drawn on the box. It may also be done by taking four readings of the azimuth circle of the instrument, with the inclination-needle adjusted to a dip of 90°, as follows : 1st, with the circle facing south (magnetic) and needle (marked side) facing south ; 2d, circle south, needle north ; 3d, circle north, needle south ; 4th, circle north, needle north. The mean azimuth circle reading for these four positions, $\pm$ 90°, gives the circle reading for the true meridian. The line upon which the setting compass is adjusted may be drawn on the box after the meridian has been thus determined by the second method, the dip-needle having, of course, been first removed. The results of the two methods should be occasionally compared.

The vertical circle being thus set in the magnetic meridian, with its divided side east, and with the marked side of the needle (face) east, two or three readings of the two ends of the needle are taken, the needle being slightly lifted so that the readings are independent. The needle is then reversed in its bearings so that it faces west, and readings are again taken as before. The circle is then reversed, so that the position is then " circle west—face east." After taking readings in this position the needle is again reversed in its bearings, the position being " circle west—face west." The polarity of the needle is then reversed, and the above readings are again taken, beginning with the last position. The table which follows will show the method of recording and reducing in order that the instrumental errors may be investigated. It will be observed that, in the case shown in the table, the centre of gravity of the needle is not in the axis of the needle, but slightly displaced towards the marked end. The errors of eccentricity are corrected by the reversals described, but they should always be reduced to a minimum.

MAGNETIC INCLINATION AT JEFFERSON CITY, MO.

Orchard of Phil. E. Chappell. Aug. 12, 1879. Instrument used, Barrow No. 9. Needle No. 3. Observer, F. E. N.

POLARITY OF MARKED END—SOUTH.							
CIRCLE EAST.				CIRCLE WEST.			
Face east.		Face west.		Face east.		Face west.	
S.	N.	S.	N.	S.	N.	S.	N.
65° 56′	66° 01′	71° 02′	71° 00′	65° 36′	65° 40′	70° 50′	70° 48′
42	65 46	70 56	70 57	33	40	71 03	71 00
40	47	71 00	71 00	35	41	03	71 00
65° 46′.0	65° 51′.3	70° 59 .3	70° 59′.0	65° 34′.6	65° 40′.3	70° 58 .6	70° 56′.0
65° 48′.6		70° 59′.2		65° 37′.4		70° 57′.3	
68° 23′.9				68° 17′.4			
68° 20′.6							

POLARITY OF MARKED END—NORTH.							
CIRCLE WEST.				CIRCLE EAST.			
Face west.		Face east.		Face west.		Face east.	
S.	N.	S.	N.	S.	N.	S.	N.
66° 56′	67° 02′	72° 22′	72° 22′	66° 51′	66° 58′	72° 23′	72° 24′
52	66 59	12	13	52	59	20	22
47	53	15	17	54	59	22	22
66° 51′.6	66° 58′.0	72° 16′.3	72° 17′.3	66° 52′.3	66° 58′.6	72° 21′.6	72° 22′.6
66° 54′.8		72° 16′.8		66° 55′.4		72° 22′.1	
69° 35′.8				69° 38′.7			
69° 37′.3							
						Resulting Inclination, 68° 58′.9	

Time of beginning, 11ʰ 15 A.M. Time of ending, 11ʰ 45ᵐ A.M. Magnetic meridian reads 16° 24′, set by compass. Left series, 68° 59′.8. Right series, 68° 58′.0.

It is best to reverse the polarity of the needle before making the first determination at any station. This is done by placing the needle upon its side upon a block, into which it fits with its upper side nearly flush with the surface of the block. A hole in the centre of the depression serves to admit the axle. Two bar-magnets are used in magnetizing the needle. Opposite poles are brought down upon the centre of the needle on opposite sides of the axle, the magnets being inclined to an angle of 40° to 45° with the horizontal plane. Preserving this inclination, the magnets are simultaneously moved in opposite directions until they leave the needle. The magnets are then lifted several inches and brought down as before at the axle ; but they should not be allowed to touch each other. The stroke should be made with uniform speed. The supporting block should have a raised guide along one side, so that the strokes may be parallel to the geometrical axis of the needle, in order to avoid eccentricity in the position of the magnetic axis. For the first magnetization at any station three strokes may be made on each side of the magnet. For the subsequent reversal at the station four strokes may be made. The magnetism of the needle diminishes somewhat as the result of the shocks incident to travel, and this is the reason for the difference in the number of strokes. After magnetization the needle should be at once placed in position ; but it should not be used for about ten minutes, as the position of the magnetic axis is likely to fluctuate, and will give very discordant results. In the table of reductions it will be observed that the positions in the left half are reproduced in the right half, so that the two means should agree, independently of the instrumental errors.

PENDULUM VIBRATIONS.

The determination of magnetic intensity or the strength of the earth's magnetic field is made by a magnet which is oscillated as a magnetic pendulum. It therefore becomes necessary to give an exposition of pendulum vibrations, in order to make the subject intelligible. It will further simplify the treatment if the ordinary gravitation pendulum is first discussed, since the unit of force is usually defined in terms of the weight of a given mass of matter. By (4) the weight of a pound at any point is g. This is the force on a pound at any point in the earth's gravitation field. It may also be called the strength of the earth's field at the point. In a study of the earth's gravitation field it is therefore neces-

Fig. 1

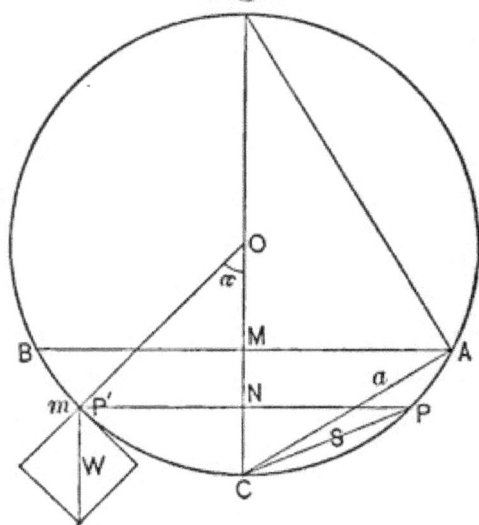

sary to find the value of g at a sufficient number of points. The direction of the lines of force are at once indicated by a plumb-line.

A simple pendulum is a heavy particle having m units of mass, suspended by a line without mass to a fixed support. If deflected, it tends to fall, and is constrained to slide in a circle about the point of suspension as centre. The particle slides down an inclined plane the angle of which continually changes, being always α, where α is the angle of deflection. The action of the field on the pendulum in a vertical direction is $w = mg$. The component along the path at any point is $mg \sin \alpha$, the acceleration at the point being $g \sin \alpha$. The velocity acquired in falling from A to P is that due to MN, the vertical height of A above P; hence

$$v^2 = 2g\, MN. \tag{7}$$

Calling L the radius of the circle or the length of the pendulum, and denoting the chord CP by s and the chord CA by a, we have, by similar triangles,

$$s^2 = 2L.\, NC$$

$$a^2 = 2L.\, MC,$$

hence

$$2L.\, MN = a^2 - s^2,$$

and by (7)

$$v^2 = \frac{g}{L}\,(a^2 - s^2). \tag{8}$$

It is evident that v will have the same value whether the sign of s is $+$ or $-$; and for any one value of s there are two values of v, which are numerically equal but of unlike sign. Hence the velocity at P and P' will be the same, and it will be the same whether the ball is ascending or descending in its path. A and B being the extremes of the excursion, the velocity is zero at those points, since $s = a$. At C the velocity is greatest, since $s = 0$. The velocity at C is

$$V = a \sqrt{\frac{g}{L}}. \tag{9}$$

The time of vibration—that is, the time required to traverse the arc AB—can be easily obtained when the arc is small so that it does not sensibly differ from its chord.

Thus, let AB represent the arc ACB of the previous figure. With C as a centre and AC, or a, as a radius, describe a circle, and suppose a point to traverse this circle with an uniform velocity of

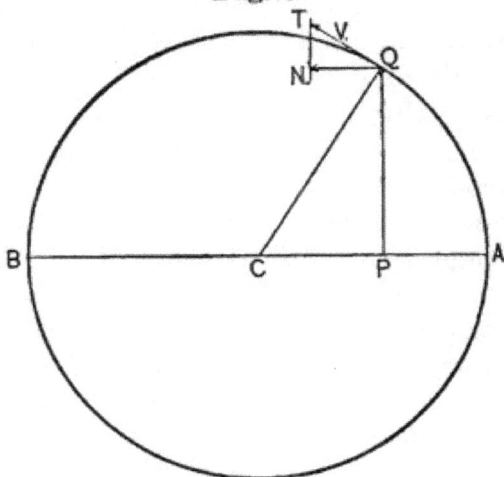

Fig. 2

$$V = a\sqrt{\frac{g}{L}}.$$

At any instant let the point be at Q. Its velocity resolved parallel to AB will be $V \cos TQN = V \cos CQP$

$$= \sqrt{\frac{g}{L}}.PQ = \sqrt{\frac{g}{L}(a^2 - s^2)},$$

since $CP = s$ and $CQ = a$. █

This is the same velocity which the pendulum has at P. Hence the point will move around the arc AQB in the same time which the pendulum requires to oscillate from A to B. Since the point moves over the arc $AQB = \pi a$ with the uniform velocity of $a\sqrt{\frac{g}{L}}$, it follows that the time required, which is the time of vibration of the pendulum, is

$$t = \frac{\pi a}{a\sqrt{\frac{g}{L}}} = \pi\sqrt{\frac{L}{g}}. \tag{10}$$

This formula will give the time with very considerable precision, if the arc AB is not over 3° or 4°.

Formula (10) applies directly to the simple pendulum, but it also holds for the compound pendulum. It, however, becomes necessary to define what is meant by the length of the pendulum, since the particles which compose it are not at equal distances from the axis of suspension. The particles nearest to the axis of suspension tend to oscillate in a shorter time than those near the lower extremity. Hence points near the axis have a longer time, and those near the bottom a shorter time, than they would have if vibrating alone around the same axis. Hence a series of points must exist which vibrate precisely as they would if they were unconnected with the system. These points are all in a straight line parallel to the axis of suspension, and constitute what is called the axis of oscillation.

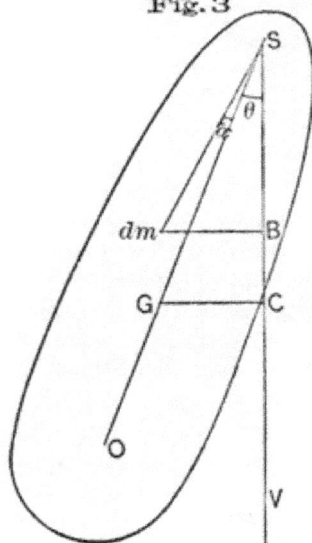

Fig. 3

The perpendicular distance between the two axes is the length of the compound pendulum. It is the length of the simple pendulum, which would make its vibration in the same time.

Let S and O represent the axes of suspension and of oscillation, G the centre of gravity of the pendulum, and dm any small element of mass.

Let $SO=L$; $SG=K$; $Sm=r$. Denote the $\angle GSC$ by θ, and $\angle mSC$ by $(\theta + \alpha)$. The planes of these angles are at right angles to the axes O and S, and, in general, do not coincide.

The pendulum might be supposed condensed upon

the vertical plane containing G, and at right angles to the axes S and O. If this change comes about by moving the particles horizontally, a thin plate will result which will have the same properties as the original pendulum. The planes of the angles θ and $\theta + \alpha$ would then coincide.

At any instant the linear acceleration of O is $g \sin \theta$. The angular acceleration of O and of every other point of the system is $\frac{g}{L} \sin \theta$. If the element dm were disconnected from the system its linear acceleration at this instant would be $g \sin (\theta + \alpha)$. The force required to produce this acceleration on dm is

$$F = dmg \sin (\theta + \alpha). \tag{11}$$

When connected with the system the real linear acceleration is $\frac{r}{L} g \sin \theta$. The force required to produce this acceleration is

$$F' = dm \frac{r}{L} g \sin \theta. \tag{12}$$

The difference $F' - F$ is a force which must be applied to dm in excess of its tangential weight component, in order to give it its actual acceleration as part of the system. The moment of this force about S is

$$r (F' - F) = \frac{r^2}{L} g \sin \theta \, dm - g \sin (\theta + \alpha) \, dm.$$

The integral of this expression for the entire pendulum is necessarily zero; hence

$$\frac{g}{L} \sin \theta \int dm \, r^2 = \int dm \, g . r \sin (\theta + \alpha).$$

In this expression the integral in the first member is the

moment of inertia I of the entire pendulum, or $M\lambda^2$, where λ is the radius of gyration. The integral in the second member is the moment of the weight of the entire pendulum, which is the same as though the whole mass were at the centre of gravity. Hence the last equation becomes

$$\frac{g}{L}\sin\theta I = MgK\sin\theta,$$

or
$$L = \frac{I}{MK}. \tag{13}$$

: The fact that $g.\sin\theta$, the real linear acceleration of O, cancels from this expression, shows that (13) is true, independent of the position of the pendulum.

The time of vibration of a compound pendulum is then found by substituting this value of L in (10), from which

$$t = \pi\sqrt{\frac{I}{MgK}}. \tag{14}$$

The denominator of (14) is the moment of the force tending to produce rotation when the pendulum is deflected 90° from its position of repose. The line through the centre of gravity, and at right angles to the two axes, is then at right angles to the lines of force of the earth's gravitation field. It is evident that some point must exist in the pendulum, at which if its entire mass were concentrated the moment of inertia about S would remain unchanged. This point is called the centre of gyration, and its distance from S is called the radius of gyration. Denoting this radius by λ, the value of I becomes $I = M\lambda^2$. Hence by (13) the relation between λ, L, and K is

$$\lambda^2 = L.K, \tag{15}$$

or the radius of gyration is a mean proportional between the distances of the centre of gravity and of the axis of oscillation from the axis S.

For the simple pendulum these distances are all equal.

If the compound pendulum consists of a thin rod having its axis of suspension intersecting the axis of figure at right angles, the expression for the length L will have the form

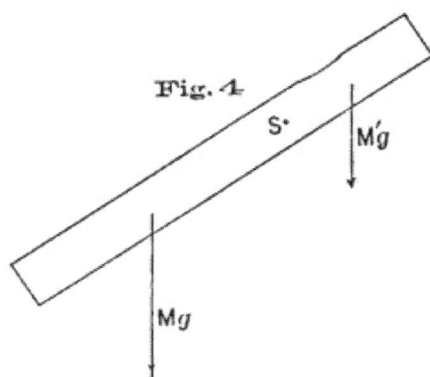

Fig. 4.

$$L = \frac{M\lambda^2 + M'\lambda'^2}{MK - M'K'},$$

the distances K and $-K$ being the distances of the centre of gravity of the two parts of the bar, separated by a plane through S and at right angles to the axis of figure. The corresponding radii of gyration are λ and $-\lambda'$, the squares of both being positive. If $MK = M'K'$, the value of L, and hence also the value of t, becomes infinite. This is the condition of a balanced lever. If the rod is of uniform section, this condition is realized when the point S is midway between the extremities of the bar. A bar of steel thus suspended will not oscillate as a gravitation pendulum, but when magnetized it will oscillate as a magnetic pendulum.

Let m represent the quantity of magnetism in each end of the magnet, measured in the units already defined (p. 11). Let F represent the action of the earth's magnetic field (or the earth-magnet) on a unit quantity of magnetism. Then the force acting on each end of the

magnet is Fm. The points of application of this force
are assumed to be at two points called poles, as the

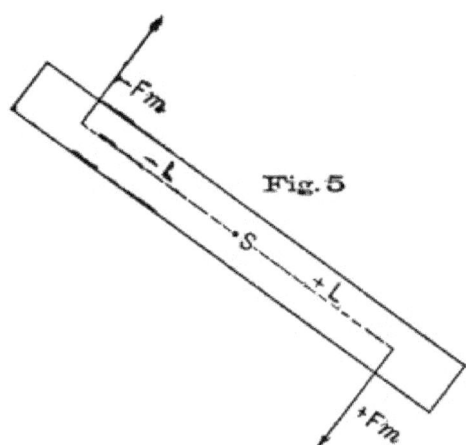

Fig. 5

points of application
in   the   gravitation
pendulum are assum-
ed to be at the centres
of gravity. The poles
are merely the cen-
tres of the forces act-
ing on the magnet.
Their position is not
related in any simple
manner to the geo-
metry of the magnet,
but depends upon the

law of magnetic distribution.   The distances of the
poles from S will therefore remain undetermined, and
may be denoted by L. The force Fm acting on each
pole of the magnet is the analogue of the force gM
acting on the centres of gravity of the pendulum. But
in the magnet the signs of m at opposite ends of the
magnet are unlike, so that the moment of the force
acting on the magnet and tending to produce rotation
when the magnet lies at right angles to the lines of
force is not zero, but $2FmL$, since the signs of m and L
reverse simultaneously. The force is applied as a couple.
In a field of unit strength this moment becomes $2mL$,
which is usually called the magnetic moment of the
magnet.   This quantity will hereafter be denoted by M.
Hence the time of vibration of a magnet free to oscillate
through the position of repose (which is the direction
of the lines of force) becomes, from (14),

$$t = \pi \sqrt{\frac{I}{FM}}. \tag{16}$$

The needle, thus suspended like a dipping-needle, measures the total force. The suspension of a needle in this manner presents great mechanical difficulties. A needle hung on a fibre of silk, and constrained to vibrate in a horizontal plane, is very much more sensitive. Such a needle determines the horizontal component of the force F, from which F is readily calculated if the inclination is known.

For the horizontal needle the time of vibration becomes

$$t = \pi \sqrt{\frac{I}{H.M}}, \qquad (17)$$

where $H = F \cos d$, in which d is the angle of inclination and H the horizontal component of the total force F.

MOMENT OF INERTIA.

The moment of inertia I can be determined by computation when the vibrating mass has a regular geometrical form, but it is usually better to use the indirect method due to Gauss. The magnet is first allowed to vibrate freely, and its time of vibration, t, is determined. The magnet is then loaded with a known mass of non-magnetic substance, so arranged with reference to the axis of vibration that its moment of inertia is known. Let it be I'.

In the first experiment the time of vibration is represented by (17). In the second experiment it is represented by the equation

$$t' = \pi \sqrt{\frac{I + I'}{H.M}} \qquad (18)$$

The observations must be corrected for torsion, and

unless the temperature of the two series is the same the
observations for the free magnet must be reduced to
that of the loaded series. The method of making these
corrections will be explained later. The observations
should be made in a room where the temperature can be
held uniform. If the determinations of t and t' are
made within a short interval of time, the values of HM
may be assumed the same in the two equations, after the
corrections for torsion and temperature have been applied. It is somewhat better to make alternating determinations of t and t', in order to eliminate changes
in H. The two equations give, when thus corrected,

$$I = \frac{t^2}{t'^2 - t^2} I' \tag{19}$$

It is most convenient to add the moment of inertia
I' in the shape of an accurately-turned ring of brass or
gun-metal, the dimensions of which are accurately determined at some known temperature. This ring is
mounted upon the magnet, the plane of the ring being
horizontal, and the axis of the ring being in the axis of
rotation. This is accomplished by first suspending the
magnet in a horizontal position, and, pointing the telescope on its scale, the adjustment of the ring must be
such as to reproduce the same pointing on the scale.

The formula for the moment of inertia of the ring
may be deduced as follows :

Calling dI' the moment of inertia of an elementary
ring of mass dm of radius r, thickness h, and radial
width dr, the value of dI' is

$$dI' = dm\, r^2 = 2\pi r . h . dr . D . r^2,$$

where D is the density of the material of which the ring
is composed. Integrating between the limits r' and r'',

$$I' = 2\pi h D \int_{r'}^{r''} r^3 dr = \frac{\pi h D}{2}(r''^4 - r'^4)$$

$$= \frac{\pi h D}{2}(r''^2 - r'^2)(r''^2 + r'^2)$$

$$= \frac{w}{2}(r''^2 + r'^2), \tag{20}$$

where w is the mass of the ring, the external and internal radii being r'' and r'. The following example will illustrate the method of finding the numerical values of I' as function of the temperature:

INERTIA RING Y OF THE U. S. COAST AND GEODETIC SURVEY.

The radii of the ring as determined by Mr. Schott were, at 62° F.,

$$r' = 1.1715 \text{ inch} = 0.09762 \text{ ft.}$$
$$r'' = 1.4219 \text{ '' } = 0.11846 \text{ ''}$$
$$w = 812.93 \text{ grains.}$$

The coefficient of expansion α was taken 0.000,010 for 1° F. For the centigrade degree it is, of course, $\frac{9}{5}$ of this quantity.

Denoting the radii at 62° F. and at θ° F. by r_{62} and r_θ, the values of the radii for a temperature θ are

$$r'_\theta = r'_{62}[1 + \alpha(\theta - 62)]$$
$$r''_\theta = r''_{62}[1 + \alpha(\theta - 62)];$$

and neglecting the square of $\alpha(\theta - 62)$, the values of r'^2_θ and r''^2_θ are

$$r'^2_\theta = r'^2_{62}[1 + 2\alpha(\theta - 62)]$$
$$r''^2_\theta = r''^2_{62}[1 + 2\alpha(\theta - 62)].$$

The errors due to these approximations at a temperature of 100° F. are about one-thirtieth of the error of measurement of the radii, assuming that the lengths in decimals of a foot are correct to the fifth decimal place.

Substituting these values in equation (20), it becomes

$$I'_\theta = \frac{w}{2} \left(r''^2_{62} + r'^2_{62} \right) [1 + 2\alpha (\theta - 62)],$$

which for computation would be put in the form

$$I'_\theta = \frac{w}{2} \left(r''^2_{62} + r'^2_{62} \right) (1 - 124\alpha) + \alpha w \left(r''^2_{62} + r'^2_{62} \right) \theta,$$

from which the following table has been computed.

MOMENT OF INERTIA OF RING Y.

FOOT-GRAIN UNITS.

$\theta°$ F	I'_θ	log. I'_θ
60	9.5771	0.98123
70	9.5790	0.98132
80	9.5810	0.98141
90	9.5829	0.98150
100	9.5848	0.98158

Degrees.	p. p. log.
1°	1
2	2
3	3
4	3
5	4
6	5
7	6
8	7
9	8

MOMENT OF INERTIA OF THE MAGNET.

The magnet being oscillated at a temperature θ, its time of vibration t becomes

$$t = \pi \sqrt{\frac{I_\theta}{H M_\theta}}. \tag{21}$$

When loaded with the inertia ring, and at a temperature θ' for magnet and ring, the time of vibration will be

$$t' = \pi \sqrt{\frac{I_{\theta'} + I'_{\theta'}}{H M_{\theta'}}}. \tag{22}$$

The thermometer should be placed within the magnet-box and read through a window in the same. The magnetic moment varies with the temperature, and therefore the value of t must be corrected to the temperature θ'. Both t and t' should also be corrected for the torsion of the suspension-fibre. The manner of making these corrections is explained later. This being done, the two values of HM being assumed equal in (21) and (22), the result is

$$I_{\theta'} \left(\frac{I_\theta}{I_{\theta'}} t'^2 - t^2 \right) = t^2 I'_{\theta'}.$$

If θ and θ' do not differ more than one or two degrees, the value of $\frac{I_\theta}{I_{\theta'}}$ may be taken as unity. The value of $I_{\theta'}$ is then

$$I_{\theta'} = \frac{t^2}{t'^2 - t^2} I'_{\theta'}. \tag{23}$$

If θ' is the higher temperature, it will be observed that the effect of calling $\frac{I_\theta}{I_{\theta'}}$ unity is to make the resulting value of $I_{\theta'}$ too small. If this difference is thought

to be appreciable, it may be assumed that the deter-
mined value corresponds to a temperature $\frac{\theta + \theta'}{2}$. But
it is very easy to make the experiments at a suffi-
ciently constant temperature, so that no correction is
needed. This is best done by making the experiments
in some dry basement-room from which artificial heat
is excluded, making use of temperature fluctuations
due to changes of temperature of the external atmo-
sphere. The magnet and ring should be continually on or in the magnetometer with the thermometer,
and the magnetometer-box should be left open during
the intervals between the observations. In handling
the ring and magnet the fingers should be covered with
non-conducting material, as rubber or rubber-cloth. If
artificial heat is used the temperature changes should
be slow, and the desired temperature should be maintained constant for some hours before the observation,
in order to allow the magnetic condition of the magnet
to become stable. The temperature should not be
raised above the highest summer heats (shade) which
are to be experienced in summer work.

At least twenty determinations of I_θ should be thus
obtained at varying temperatures θ. As the change in I
is simply due to expansion, and is very small, the function may be considered a linear one, and may be represented by the equation

$$\log I = \log I + (\theta - \theta)\, \varDelta I, \qquad (24)$$

where $\log I_0$ is the value of $\log I$ at a temperature θ_0,
the mean temperature of the series, and $\varDelta I$ is the change
in $\log I$ per degree of change in temperature. The
value of $\log I_0$ is the mean of the values $\log I_\theta$ given in

the third column of the following table. It then remains to find $\varDelta I$, for which the 23 observations furnish
23 equations. The value of $\varDelta I$ may be determined by
graphical methods. The simultaneous values of log I_θ
and θ are plotted, and by aid of a thread the position
of the line represented by (24) is determined, the constants for which are thus easily determined by well-
known methods of analytical geometry. The constants
may also be determined by means of the method of
least squares. When the computations are properly
arranged this involves little labor, and the calculation is
made in the table as an example of the method. The
discussion of this method is given in the appendix.

Let it be required to assign a value to $\varDelta I$ in order to
most nearly satisfy the 23 equations. If any value at
random be assigned to $\varDelta I$, then in general the value
of log I_0 – log $I_\theta + (\theta - \theta_0)\, \varDelta I$ will not be zero. If
its value be denoted by e, and if for convenience we put

$$y = \log I_0 - \log I_\theta$$

and

$$u = \theta - \theta_0,$$

there will be 23 equations of the form

$$e = y + u\, \varDelta I.$$

The values of e should all be small numerically, some
being minus and some plus. The sum of the values e^2
or Σe^2 for the twenty-three equations should be a minimum. As is shown in the appendix, the value of $\varDelta I$
which will make Σe^2 a minimum is

$$\varDelta I = -\frac{\Sigma\, uy}{\Sigma u^2}$$

where $\Sigma\, uy$ is the sum of the twenty-three products of
the simultaneous values of u and y, Σu^2 being the sum

of the values u^2. This computation is given in the final columns of the table. The resulting value is

$$\varDelta I = + \frac{0.12511}{5938.9} = 0.000021$$

and hence

$$\log I_\theta = 1.22294 + 0.000021 \, (\theta - 64.8), \qquad (25)$$

from which the small table for the log moment of inertia of the magnet at various temperatures has been calculated.

COMPUTATION FOR MOMENT OF INERTIA OF MAGNET C_6.

Date.	θ	Log. I_θ	y	u	uy	u^2
1880–81						
Oct. 18	68.7	1.22326	−0.00032	+ 3.9	−0.00125	15.21
Nov. 3	65.5	1.22295	+ 1	+ 0.7	+0.00001	0.49
" 10	58.4	1.22587	− 293	− 6.4	+0.01875	40.96
Dec. 8	65.0	1.22535	− 241	+ 0.2	−0.00048	0.04
" 31	54 0	1.22465	− 171	−10.8	+0.01847	116.64
Jan. 1	48.5	1.22238	+ 56	−16.3	−0.00913	265.69
" 2	51.0	1.22239	+ 55	−13.8	−0.00759	190.44
" 3	53.0	1.22227	+ 67	−11.8	−0.00791	139.24
" 4	51.1	1.22126	+ 168	−13.7	−0.02302	187.69
" 4	50.4	1.22080	+ 214	−14.4	−0.03082	207.36
" 13	63.0	1.22540	− 247	− 1.8	+0.00445	3.24
" 17	52.0	1.22260	+ 34	−12.8	−0.00435	163.84
" 18	48.0	1.22135	+ 159	−16.8	−0.02671	282.24
" 19	53.5	1.22248	+ 46	−11.3	−0.00520	127.69
" 26	66.5	1.22346	− 52	+ 1.7	−0.00088	2.89
" 28	61.5	1.22188	+ 106	− 3.3	−0.00350	10.89
Feb. 2	61.2	1.22241	+ 53	− 3.6	−0.00191	12.96
" 4	65.5	1.22223	+ 71	+ 0.7	+0.00050	0.49
" 5	68.1	1.22177	+ 117	+ 3.3	+0.00386	10.89
" 23	99.7	1.22511	− 217	+34.9	−0.07573	1218.01
" 26	103.5	1.22224	+ 70	+38.7	+0.02709	1497.69
March 4	96.0	1.22138	+ 156	+31.2	+0.02867	973.44
" 29	86.5	1.22425	− 131	+21.7	−0.02843	470.89
Means..	64.8 θ_0	1.22294 $\log I_0$	....	....	−0.12511 Σuy	5938.92 Σu^2

The following table for the log of the moment of inertia has been calculated from (25). The table of proportional parts gives the differences from 1° to 9°. For instance, if this magnet were oscillated at a temperature 87°, the log of the moment of inertia of the magnet would be 1.22341.

MOMENT OF INERTIA OF MAGNET C_6, FROM (25).

θ	Log. I_θ
50°	1.22263
60°	1.22284
70°	1.22305
80°	1.22326
90°	1.22347
100°	1.22368

Degrees.	p. p. log.
1	2
2	4
3	6
4	8
5	11
6	13
7	15
8	17
9	19

CORRECTION OF THE OSCILLATION SERIES FOR TORSION.

If a brass weight having the same moment of inertia as the magnet were suspended on the silk fibre, the torsion of the fibre would cause it to oscillate slowly. The time of vibration would be expressed by an equation of the form (14) or (16), viz.:

$$t'' = \pi \sqrt{\frac{I}{\delta}}, \tag{26}$$

where δ, the directive constant depends in this case upon the length, diameter, and material of the fibre. A horizontal magnet having the same value of I would, therefore, have the time of vibration,

$$t' = \pi \sqrt{\frac{I}{HM + \delta}}. \tag{27}$$

If the magnet were not affected by torsion its time of vibration would be

$$t = \pi \sqrt{\frac{I}{HM}}; \tag{28}$$

hence from (27) and (28)

$$\frac{t^2}{t'^2} = 1 + \frac{\delta}{HM}. \tag{29}$$

It is here assumed that the line of detorsion lies in the magnetic meridian, so that the fibre does not influence the position of the needle when it is at rest.

In order to find the value of $\dfrac{\delta}{HM}$, turn the torsion-head, to which the upper extremity of the fibre is attached, through, say, 90°; the magnet will follow through an angle v, where v will be usually about six to ten minutes. The number of minutes of twist in the fibre will be 5400 — v. Since v is small, the moment of the earth's force tending to bring the magnet into the meridian will be HMv (really $HM \sin v$); the moment of the force of torsion of the fibre tending to deflect the magnet still more is $(5400 — v)\, \delta$. Hence for equilibrium

$$(5400 — v)\, \delta = v\, HM,$$

or

$$\frac{\delta}{HM} = \frac{v}{5400 — v}.$$

This in (28) gives, since $\dfrac{v}{5400}$ is small,

$$t^2 = \frac{5400 + v}{5400}\, t'^2. \tag{30}$$

This gives the necessary correction when v is measured. In determining v the torsion-head is first turned 90° in a + direction, and then backwards in a — direction 180°, and then again in a + direction 90°, which should reproduce the original scale-reading. The differences between the scale-readings should correspond to v, $2v$, and v, which, added together and divided by four, gives the value of v. If the twist is not all removed from the fibre, the differences for the + and the — position will not be equal. The observations and computation for torsion will be found in a specimen table of oscillations, p. 46.

TEMPERATURE CORRECTION FOR MAGNETIC MOMENT.

The magnetic moment of a magnet diminishes slightly for an increase of temperature, and the change may be assumed to take place in accordance with the equation:

$$M_{\theta'} = M_\theta \left[1 - q \left(\theta' - \theta \right) \right], \tag{31}$$

where q is a very small quantity, representing the fraction of itself by which M_θ diminishes when heated through one degree. The corrections heretofore discussed having been made, the times of vibration at the two temperatures will be

$$t = \pi \sqrt{\frac{I}{H M_\theta}} \tag{32}$$

$$t' = \pi \sqrt{\frac{I}{H M_{\theta'}}}. \tag{33}$$

Dividing one of these equations by the other, and replacing $M_{\theta'}$ by its value in (31), the resulting equation is:

$$t^2 = \left[1 - q \left(\theta' - \theta \right) \right] t'^2, \tag{34}$$

which gives the time of vibration of the magnet at a temperature θ, if the time of vibration at θ' is known, and in terms of the coefficient q. The method of determining this quantity will be explained later. It should be observed that the temperature correction is by far the most difficult and important of the corrections to be made in magnetic observations. In fieldwork it is particularly troublesome. It is better in such cases to use a tent with double walls and roof or with a

large fly, so as to keep the sunlight from the inner tent.
It should be open on opposite sides, in order to admit
free circulation of air, although it is not good to have
the magnet-box too much exposed to wind.   In the deflection observations, to be discussed later, the temperature of the deflecting magnet must be carefully determined.

The formula for oscillations becomes, therefore,

$$HM = \frac{\pi^2 I}{t^2},\qquad(35)$$

where

$$t^2 = \left(\frac{5400 + v'}{5400}\right)[1 - (\theta' - \theta)q]\,t'^2.\quad(36)$$

The temperature θ is usually that of the deflection
series.

The value I in (35) is of course $I_{\theta'}$, or the moment of
inertia at the temperature θ' at which the magnet was
actually oscillated.   This is easily shown by assuming as
sufficiently precise, the equation

$$I_{\theta'} = I_\theta [1 + c(\theta' - \theta)],$$

and giving the values I_θ and $I_{\theta'}$ to this quantity in equations (32) and (33).   The proof easily follows.

Dept. Mech. Eng.

HORIZONTAL INTENSITY—OSCILLATIONS.

Date Aug. 12, 1879.   Station, Jefferson City, Mo., Chappell's orchard.   Instrument, University Magnetometer, Magnet C_6. Watch, Jürgensen No. 10890 ; loses 3^s per day.   Observers, F. E. N. and J. W. S.

No. Oscill.	Watch reads.	Temp. θ'	Extreme scale-readings.	Time of 100 oscill.
		81°	$70^d.5 - 90^d.0$	
0	$9^h \ 13^m \ 41^s.5$		$10\ osc.=68^s.2$	
10	14 49 .7		68 .3	
20	15 58 .0		68 .1	
30	17 06 .2		68 .1	
40	18 14 .3		68 .3	
50	19 22 .6		Mean 68 .2	
			$100\ oscill.=$	
100	25 04.0	81°	$11^m \ 22^s.0$	$11^m \ 22^s.5$
110	26 12.2			11 22 .5
120	27 20.7			11 22 .7
130	28 29.0			11 22 .8
140	29 37.0			11 22 .7
150	30 45.2			11 22 .6
		82°	$74^d.6 - 86^d.8$	
Means	θ'	81.3		11 22 .63

1 scale div. = 2′.345.

Torsion circle.	Scale.		Mean.	Diff.		Logs.
180°	74.6	86.8	80.7	1.7	$v = 4'.1$	3.73272
+90	68.1	90.0	79.0		$5400 + v$	
−180	74.2	91.0	82.6	3.6		6.26761
+90	71.9	90.0	80.9	1.7	5400 (a.c.)	
				$\overline{7.0}$		
Mean v = 1.75 div.					Correction	0.00033

HORIZONTAL INTENSITY—COMPUTATION.

$$t^2 = t'^2 \left(\frac{5400 + v'}{5400}\right)[1 - (\theta' - \theta)q].$$

Observed time of 100 oscill...................... 682ˢ.63
Time " 1 " .... 6 .8263
Correction for rate.................... +0 .0002
$t' =$ 6 .8265

q	0.00048		Logs.
$\theta' - \theta$	+4.1	t'	0.83419
$q(\theta' - \theta)$	0.00197	t'^2	1.66838
$1 - q(\theta' - \theta)$	0.99803	$\frac{5400 + v'}{5400}$	0.00033
		$1 - (\theta' - \theta)q$	9.99914
		t^2	1.66785
$MH = \frac{\pi^2 I}{t^2}$		a.c. t^2	8.33215
		π^2	0.99430
		I	1.22340
$M = 0.7440$		MH	0.54985
$H = 4.767$		M	9.87158
		H	0.67827

Deflection series, 12th Aug., at 8.45 A.M.
$\theta = 77.2.$

I is from an old table, and was correct at that date for temp. of $\theta' = 81.3$.	$\frac{M}{H}$	9.19331
	MH	0.54985
	M^2	9.74316
	M	9.87158

REDUCTION OF THE TIME OF OSCILLATION TO THAT OF AN INFINITELY SMALL ARC.

The time of vibration for an infinitely small arc has already been shown to be

$$t_0 = \pi \sqrt{\frac{I}{HM}}.$$

The complete time of vibration of any oscillating body in which the force which draws it towards its position of repose is proportional to the sine of the angle of displacement, is

$$t = t_0 \left(1 + \tfrac{1}{4}\sin^2 \frac{\alpha}{4} + \frac{9}{64}\sin^4 \frac{\alpha}{4} \cdots \right)$$

where α is the total arc described. This formula or its equivalent can be found in any good treatise on analytical mechanics.

The formula may be written

$$t_0 = t - \left(\tfrac{1}{4}\sin^2\frac{\alpha}{4} + \frac{9}{64}\sin^4 \frac{\alpha}{4} \right) t,$$

where t is the observed time of vibration. The value of the expression within the parenthesis, denoted by A, is given in the table.

It is never necessary with a collimator magnet, or with one employing a reflecting scale according to the original method of Gauss, to make the value of α over 40' to 1°, so that this correction may ordinarily be neglected altogether.

In case the correction is to be made, the first term of the series only need

α	A
0°	0.00000
1	00
2	02
3	04
4	08
5	12
6	17
7	23
8	30
9	39
10	0.00048

be used, and the sine may be replaced by the arc. The formula then becomes

$$t' = t\left(1 - \frac{\alpha^2}{64}\right),$$

where α is the mean of the arcs of the first and last oscillations. An equivalent expression is

$$t_0 = t\left(1 - \frac{\alpha'\alpha''}{64}\right),$$

where α' and α'' are the arcs of the first and last oscillations. The observations should be so arranged that the amplitude does not diminish more than one-third during the determination in order to apply these latter formulæ.

In order to obtain the corrected value t^2 in equation (36), if correction is to be made for amplitude, as well as torsion and temperature, the latter formula becomes

$$t^2 = \left(\frac{5400 + v}{5400}\right)[1 - (\theta' - \theta)q]\left(1 - \frac{\alpha'\alpha''}{32}\right)t'^2, \quad (37)$$

where the arcs are, of course, expressed in circular measure.

In the example given in the blank the person observing the magnet-scale called "time" on every tenth oscillation, and the assistant read the watch, observing the second-hand by means of a magnifying-glass. The time may also be taken by a single observer, if provided with a chronometer having a jumping second or half-second hand. The beat is taken up and carried in mind just before the oscillation to be timed occurs. The second is best divided by noting the position of the middle division of the scale at the beats before and after its transit of the cross-hair. It is not necessary to count the oscillations between 50 and 100, as the time

of occurrence of the hundredth beat can be calculated,
with a precision sufficient to enable one to recognize it
with certainty, after the first half of the series is made,
as is shown in the blank, where the time of 100 oscillations, calculated from the first half of the series, is
$11^m 22^s.0$, which, added to the watch-reading of the
initial observation, $9^h 13^m 41^s.5$, gives $9^h 25^m 03^s.5$ as
the calculated time of occurrence of the hundredth oscillation. The observation for time should, of course,
be made when the magnet is at the middle of its swing.

Formula (35) gives the value of H in terms of the unknown quantity M. If this quantity does not change in
time, (35) may be used in making relative determinations. This method may be used with an old magnet,
if it is protected from the influence of other magnets
and from mechanical shocks. The slight decrease in
the value of M which may be expected may be de-
tected by oscillating before and after the tour, at a base
station, the change being assumed constant.   It is,
however, always better to determine the value of M
fifteen or twenty times in the course of a summer.   This
is done by obtaining an independent equation involving
M and H, as will now be shown.   This method is due
to Gauss.

DEFLECTION SERIES FOR INTENSITY.

The magnet used in the oscillation series is replaced by a small magnet,

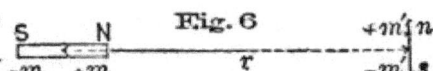

Fig. 6

$n\ s$, Fig. 6. The oscillation or "intensity" magnet, as it is sometimes called, is then mounted on a horizontal bar, with its longitudinal axis at right angles to the plane of the magnetic meridian, the prolongation of its axis bisecting the axis of the suspended needle, as is shown in Fig. 6. This causes a deflection of the needle, the angle of deflection (for a given distance r between the centres of the magnets) being greater as the value of H is less.

In some magnetometers the telescope and magnet-box are mounted on a common azimuth circle, and the whole instrument turns about the vertical axis of the circle. The telescope is always adjusted on the central scale-division. The angle of deflection is thus read on the azimuth circle. In others the instrument remains in position in the plane of the magnetic meridian, the angle of deflection being read on the magnet-scale. The latter form was that originally devised by Gauss. These instruments bear to each other the same relation as the sine and the tangent galvanometer. The discussion for both forms of instruments will be given.

Let m be the strength of the poles of intensity magnet $N\ S$, $2l$ the distance between its poles, and let m' and $2l'$ be the analogous values for the needle $n\ s$. Let r be the distance between the centres of the magnets. The distance l' being small compared with r, the repulsion of N on n will be approximately $\dfrac{mm'}{(r-l)^2}$,

and the attraction of S on n will be $\dfrac{mm'}{(r+l)^2}$.   The difference between these expressions will approximately repre-
sent the resultant repulsion of NS on n, or

$$F = mm' \left[\frac{1}{(r-l)^2} - \frac{1}{(r+l)^2} \right] = mm' \frac{4rl}{(r^2-l^2)^2}$$

$$= 2Mm' \frac{r}{(r^2-l^2)^2} = \frac{2Mm'}{r^3} \left[1 + 2\frac{l^2}{r^2} + \cdots \right]$$

since higher powers of $\dfrac{l}{r}$ may be omitted.   Hence the
moment of the couple acting on ns is

$$2l'F = \frac{2MM'}{r^3} \left(1 + 2\frac{l^2}{r^2} + \cdots \right).$$

The needle is deflected through a
small angle, u, and comes to rest when
the deflecting moments of the magnet
NS and the earth's field are equal.
The moment of the force due to the
earth's field is $M'H \sin u$.   The force
of NS on ns being, for the tangent
magnetometer, assumed to act at right
angles to the magnetic meridian, the
moment of this force will be

$$2l'F \cos u.$$

For equilibrium

$M'H \sin u$

$$= \frac{2MM'}{r^3} \left[1 + 2\frac{l^2}{r^2} + \cdots \right] \cos u.$$

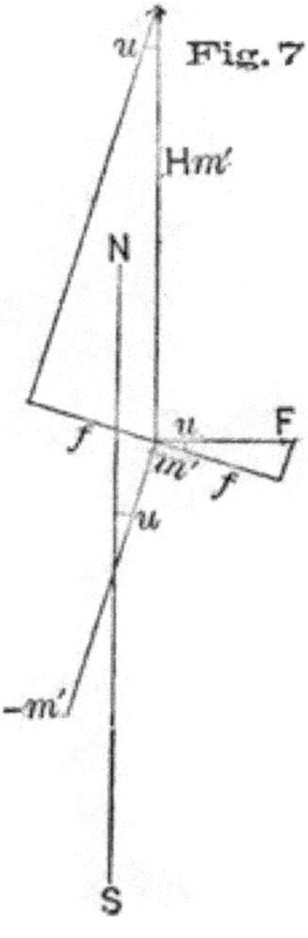

Fig. 7

This equation involves several approximations not made in the great discussion of Gauss. The determinations are, however, subject to errors of adjustment and to errors dependent on the position of the magnetic axes of the magnets, which do not coincide with the geometrical axes. In practice, therefore, the value of $2l^2$ in the small term $2\dfrac{l^2}{r^2}$ is replaced by a constant, P, and the errors of adjustment and of the approximations are thrown upon this quantity. The equation then becomes

$$\frac{M}{H} = \tfrac{1}{2}r^3 \tan u \left[1 - \frac{P}{r^2} + \ldots\right], \qquad (38)$$

which is the same result as is obtained by the more general discussion of Gauss.

At each station the value of u is observed for two distances, r and r'. The two equations thus obtained can be combined by eliminating $\dfrac{M}{H}$ if the two determinations are made near together in time, and thus a value of P is obtained which will satisfy the two equations. The residual errors of adjustment and of eccentricity are thus thrown upon the quantity P, which always turns out to be small, if r and r' are properly chosen. The sign of P is sometimes $+$ and sometimes $-$. For obtaining an average value of P it is customary to make at least twenty observations, and for reducing the observations of a season it is sufficient to take an average value of P as determined at all the stations.

When the sine magnetometer is used the deflecting magnet is at right angles to the needle when the reading is taken. The moment of the deflecting magnet on the needle is then $2l'F$ instead of $2l'F \cos u$. Hence the final equation becomes

$$\frac{M}{H} = \tfrac{1}{2}r^3 \sin u \left[1 - \frac{P}{r^3} + \ldots \right]. \qquad (39)$$

If the two deflection series are made at different temperatures they must be reduced to a common temperature. Let $\theta'' =$ the mean temperature of one series, and θ that of the other. Let u' be the observed angle of deflection in the first series : it is required to find the angle u of deflection if the temperature had been θ, or that of the other series. For the tangent instrument the two equations become

$$\frac{M_{\theta''}}{H} = \tfrac{1}{2}r^3 \tan u' \left[1 - \frac{P}{r^3} + \ldots \right]$$

$$\frac{M_{\theta}}{H} = \tfrac{1}{2}r^3 \tan u \left[1 - \frac{P}{r^3} + \ldots \right].$$

Dividing one equation by the other, and reducing by equation (31), the result is

$$\tan u = \frac{\tan u'}{1 - q\,(\theta'' - \theta)}. \qquad (40)$$

The oscillation series is then also reduced to a temperature θ, as has been already explained.

If θ and θ'' do not differ more than two or three degrees, the mean of the two values $\frac{M}{H}$ unreduced for temperature may be taken for a temperature $\tfrac{1}{2}(\theta + \theta'')$, to which the oscillation series is then reduced.

In the sine magnetometer no twist is developed in the sustaining fibre, but in the tangent instrument a torsion correction is needed, which is deduced as follows :

1. The twist having been removed from the fibre by

means of the brass torsion-weight, the small magnet being suspended, if the torsion-head is turned through 90° the magnet is displaced through an angle $v°$. The number of degrees of twist in the fibre is $90° - v°$. Hence 1° of twist in the fibre will displace the magnet

$$\frac{v}{90 - v}.$$

2. The initial conditions, being as in 1, deflect the magnet through an angle u' by means of some other magnet. But for the torsional effect of the fibre the deflection would be a little greater, the angle being u. Hence a twist of u' degrees in the fibre causes a displacement of $u - u'$ degrees of the magnet, or a twist of 1° displaces the magnet $\dfrac{u - u'}{u'}$ degrees.

Hence

$$\frac{v}{90 - v} = \frac{u - u'}{u'}$$

or

$$\frac{u - u' + u'}{u'} = \frac{v + 90 - v}{90 - v}$$

and

$$\frac{u}{u'} = \frac{90}{90 - v} = \frac{1}{1 - \frac{v}{90}} = 1 + \frac{v}{90}.$$

Reducing to minutes,

$$u = \frac{5400 + v}{5400} u', \tag{41}$$

where v is in minutes. The torsion correction, therefore, has the same form as in the oscillation series.

The form for the observation of the deflection series is here given. The deflecting magnet C_s is first placed on the west end of the bar, then on the east end, etc., the direction which the north end points being shown

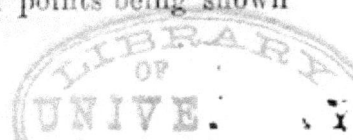

HORIZONTAL INTENSITY—DEFLECTIONS.

Date Aug. 12, 1879.   Station, Jefferson City (Chappell's). Magnet C_6 deflecting, Magnet C_{17} suspended.   Instrument, University Magnetometer.   Observer, F. E. N.

C_6.	North end.	Time.	Temp. θ	Scale-reading.	Alternate Means.	Diffs.	r.
West.	W	8^{h}29^m	75	33.0		93.60	
	E	8 33	76	126.6	33.00	93.55	
	W	8 36	76.5	33.0	126.55		
	E	8 38	76	126.5			
		8 34	75.9			93.57	$r = 2$ feet.
East.	E	8 30	75.5	125.9		92 05	
	W	8 32	76	33.8	125.85	92.15	
	E	8 35	76.5	125.8	33.65		
	W	8 40	77	33.5			
		8 34	76.2			92.10	
Means			76.0		$2u$	92.835	

Computation, $\dfrac{M}{H} = \frac{1}{2}r^3 \tan u \left(1 - \dfrac{P}{r^2}\right).$

Torsion circle.	Scale.		Mean.	Diffs.		Logs.
180	....	....	79.7	3.9	$u = 46.417$	1.66668
270	....	....	83.6	7.9	1 div. = 2'.89	0.46090
90	....	....	75.7	4.0	$\dfrac{5400 + v}{5400}$	0.00091
180	79.9	79.5	79.7			
					$u = 134'.43$ } = 2° 14'.43 }	2.12849
Mean $v =$			3.95		tan u	8.59241
					r^3	0.90309
$v = 11'.4$			Logs.		$\frac{1}{2}$	9.69897
$5400 + v$		3.73330			$1 - \dfrac{P}{r^2}$	9.99887
5400 (a.c.)		6.26761				
Correction		0.00091			$\dfrac{M}{H}$	9.19334

in the second column. The order of the experiments is indicated in the time column, and is so arranged as to eliminate the hourly change in declination. The method of reduction will be sufficiently apparent by inspection of the blank. A second series with $r = 1.75$ ft. at a mean temperature θ of 78.4 gave for a corrected value of u 3° 20'.66. These two series were used in calculating the value of P in equation (38), which was found to be $+ 0.0083$. The value of P used in the final reduction is, however, a mean for the work of a season. The resulting value of $\log \frac{M}{H}$ for the second position was 9.19326. Hence for the two series the mean temperature θ is 77°.2, and the mean value $\frac{M}{H}$ is 9.19331, which are used in the oscillation blank.

DETERMINATION OF THE TEMPERATURE COEFFICIENT q.

The temperature coefficient has been used in correcting both the oscillation and the deflection series. Either of these correction formulæ may be used in determining the value of q if all the other values in the equation are directly observed. If the oscillation series is used the magnet-box may be surrounded by a copper waterjacket, with windows, closed with double walls of mica or glass, to admit of the proper illumination and to enable one to read the thermometer, which should be placed within the magnet-box. The copper vessel must be first examined to see if it acts magnetically, and, if so, its position must remain unchanged during the determination. The jacket is first filled with ice-water,

3

and ice-water is allowed to slowly run through the
jacket in such a manner as to secure as uniform a circulation as possible.   The temperature should be held
constant for at least an hour before observations are
begun.   The temperature need not fall below 40°.
After the ordinary oscillation determination, as shown
in the blank, the ice-water is drawn off and warm water
or steam is passed through the jacket.   The change in
temperature should be gradual, and the higher temperature (maximum summer-heat in the shade) should again
be maintained for an hour.   Unless these precautions
are taken fallacious values of q will result.   If another
iustrument is not available for the simultaneous determination of H (relative determinations only are needed),
the lower temperature should again be reproduced with
similar precautions, in order to eliminate changes in H;
the mean of the two determinations at the mean of the
lower temperatures (which may differ two or three degrees) being combined with that at the higher.   Since
the magnet is oscillated at the two temperatures, the
value of I for those temperatures must be used.

The equations for the two temperatures are :

$$t^2 = \pi^2 \frac{I_\theta}{H M_\theta}$$

$$t'^2 = \pi^2 \frac{I_{\theta'}}{H' M_{\theta'}}$$

Dividing one equation by the other, and combining
with (31),

$$\frac{t^2}{t'^2} = \frac{I_\theta}{I_{\theta'}} \frac{H'}{H} [1 - (\theta' - \theta) q],$$

from which

$$q = \frac{1}{\theta' - \theta} \left(1 - \frac{t^2}{t'^2} \frac{I_{\theta'}}{I_\theta} \frac{H}{H'} \right). \qquad (42)$$

If the determinations at θ and θ' have alternated, as explained above, the ratio $\dfrac{H}{H'}$ may be assumed unity. Its value may, however, be determined by a deflection magnetometer, which may easily be extemporized in any good laboratory; and this is strongly recommended. This instrument should be in a room of constant temperature. The ratio

$$\frac{H}{H'} = \frac{\tan u'}{\tan u}$$

if the tangent magnetometer is used [eq. (38).] The angles u' and u must, however, be corrected for change in declination, which may be done by removing the deflecting magnet.

The value of q may also be determined by the method of deflections. It is also best in this method to use a second instrument to determine changes in declination. The change in H should also receive a correction as before. The deflection-magnet is put in position on the deflection-bar and surrounded with a copper water-jacket. The deflected needle and declination are simultaneously read at the lower temperature. If a second instrument is not available, the angle of deflection must be determined by removing the deflection-magnet from the bar. It is then more difficult to control its temperature, but with proper care and patience the method will give good results.

If the sine magnetometer is used the deflection formula (39) gives the equation

$$\frac{M_{\theta'}}{M_\theta} = \frac{\sin u'}{\sin u};$$

hence

$$\frac{M_{\theta'} - M_\theta}{M_\theta} = \frac{\sin u' - \sin u}{\sin u}. \tag{43}$$

By (31)

$$\frac{M_{\theta'} - M_{\theta}}{M_{\theta}} = - q\,(\theta' - \theta). \qquad (44)$$

Since u' and u differ very little from each other, $\sin u' - \sin u = \cos u\,(u' - u)$; hence, by (43) and (44),

$$q = \cot u\,\frac{u - u'}{\theta' - \theta}.$$

If $u' - u = d$, measured in scale divisions, and the value of one scale division in minutes be s, the arc of $1'$ in terms of radius being L, then the above equation becomes

$$q = \cot u\,\frac{sdL}{\theta' - \theta}, \qquad (45)$$

where u is the angle of deflection at the lower temperature θ.

If the tangent magnetometer is used equation (38) gives, as in the previous case,

$$\frac{M_{\theta'} - M_{\theta}}{M_{\theta}} = \frac{\tan u' - \tan u}{\tan u} = - q\,(\theta' - \theta).$$

If u' and u are small (about two degrees), $u - u'$ being not over two or three minutes, which is about its usual value, the numerator $\tan u' - \tan u$ may be written $u' - u$. Under these conditions the expression for q is the same as that deduced for the sine magnetometer (45). In case the above approximation is not admissible the formula becomes

$$q = \cot u\,\frac{\tan u - \tan u'}{\theta' - \theta}. \qquad (46)$$

The following practical example will show the method of determination :

Determination of q for magnet C_6 of U. S. C. and G. Survey, March 11, 1881. F. E. N., observer.

Magnet C_6 deflecting C_{17}, which is suspended in the University magnetometer, mounted on south pier of the clock-room of Washington University. Declinometer No. 3, U. S. C. and G. Survey, with magnet No. 1 suspended, was mounted on north pier, in order to correct for hourly change. Both magnets had been suspended for a week in order to render the fibres constant, torsion being corrected. The scale values of the magnets were, C_{17}, $2'.89$; No. 1, $1'.902$; and the scales are so mounted that when the easterly declination increases, the scale-reading of No. 1 increases, while that of C_{17} decreases. At 2 o'clock P.M. the scales read :

$$C_{17}, 79.0. \qquad\qquad \text{No. 1, } 77.95.$$

At 3 P.M. magnet C_6 was put in place, deflecting C_{17}, r being twenty-one inches. It was surrounded with a copper jacket of ice-water, which was fed by a drip of ice-water from a piece of ice during the night. The next morning the scales read as follows:

Hour.	SCALE-READING.		Temperature of	
	No. 1.	C_{17}.	C_6.	Room.
8^h 30^m	82.0	149.15	67.5	70
8 45	81.6	149.5	67.4	70
9 40	80.55	150.0	67.5	70
10 15	79.9	150.2	67.6	70

Mean,   67.5.

At 10.15 the ice-water was gradually removed and

hydrant-water substituted, which was gradually warmed
with a Bunsen flame.   The readings were then:

Hour.	SCALE-READING.		Temp. of C_6.	
	No. 1.	C_{17}.		
11^h 55^m	77.3	150.9	103	71
12 10	77.4	151.2	104	72
20	77.4	151.0	105	72
35	77.5	151.0	104	72
50	77.5	150.9	105	72
1 00	77.5	150.9	104	72
1 17	77.6	151.1	105.3	72

Mean,   104.3.

At 1.18 P.M. the doors of the water-bath were opened
and C_6 was allowed to cool down slowly, the water being
gradually replaced by ice-water as before.   The readings were then :

Hour.	SCALE-READING.		Temp. of C_6.	Temp. of Room.
	No. 1.	C_{17}.		
4^h 00^m	78.3	151.5	66	72
14	78.5	151.8	66	72
25	78.65	151.8	66.8	72.2
36	78.65	151.75	67.5	72.5
45	79.0	151.7	67.8	72.2

Mean,   66.8.

At 4.55, C_6 being removed, the suspended magnets read :

No. 1, 78.95. C_{17}, 79.35.

The mean reading of No. 1 was 78.7, and the readings of C_{17} were corrected to this, as is shown in the following table. In applying the correction to the scale-reading of C_{17} its sign is reversed, by reason of

	Magnet No. 1.		Correction in scale div. of C_{17}.	Reading of C_{17}.		Mean C_{17}.
	Reading.	Correction to Mean.		Observed.	Corrected.	
Series I.	82.0	−3.3	+2.2	149.15	151.4	
	81.6	−2.9	+1.9	149.5	151.4	151.25
	80.55	−1.85	+1.2	150.0	151.2	
	79.9	−1.2	+0.8	150.2	151.0	
Series II.	77.3	+1.4	−0.9	150.9	150.0	
	77.4	+1.3	−0.86	151.2	150.3	
	77.4	+1.3	−0.86	151.0	150.2	
	77.5	+1.2	−0.8	151.0	149.2	150.04
	77.5	+1.2	−0.8	150.9	150.1	
	77.5	+1.2	−0.8	150.9	150.1	
	77.6	+1.1	−0.7	151.1	150.4	
Series III.	78.3	+0.4	−0.26	151.5	151.24	
	78.5	+0.2	−0.13	151.8	151.67	
	78.65	+0.05	−0.03	151.8	151.77	151.66
	78.65	+0.05	−0.03	151.75	151.72	
	79.0	−0.3	+0.19	151.7	151.89	

the fact that the scales read in opposite directions, as stated.

For the lower temperature θ, Series I. and III., the mean readings are :

	θ	C_{17} reads
Series I.,	67.5	151.25
" III.,	66.8	151.66
Means	67.15	151.46

For the higher temperature θ',

	θ'	C_{17} reads
Series II.,	104.3	150.04

Hence a change of 37°.15 F. in the temperature of magnet C, produces a change of 1.42 scale divisions in the reading of C_{17}.

The angle of deflection u at the lower temperature is determined from the simultaneous readings before and after the experiments, as follows :

	Before Series I.	After Series III.
No. 1 read......................	77.95	78.95
Mean of series...............	78.7	78.7
Correction to mean..........	+0.75	−0.25
Reduced to scale of C_{17}......	−0.49	+0.16
C_{17} read....................	79.0	79.35
C_{17} corrected...............	78.51	79.51
C_{17} during deflection........	151.25	151.66
Angle of deflection.........	72.74 div.	72.15 div.

Hence the mean angle of deflection u at the lower temperature is 72.45 scale divisions of C_{17}, or 3° 29'.4.

The value of q is therefore computed as follows :

		Logs.
$s = 2.89$		0.461
$d = 1.42$		0.152
$L = 0.00029$		6.464
$u = 3° 29'.4$	cot.	1.215
$\theta' - \theta = 37.15$	$a.c.$	8.430
$q = 0.00047$		6.722

The observer should then compute a table of the values of $\log [1 - (\theta' - \theta) q]$ for values of $\theta' - \theta$ between $+9$ and -9, and a table of proportional parts for tenths of a degree, to facilitate the reduction of the oscillation series. The value of q depends on the nature and hardness of the steel of which the magnet is composed. In the above case a previous determination some years before gave for this magnet a value $q = 0.00048$.

SYSTEMS OF UNITS.

In the government surveys in England and America the fundamental units taken have been the foot, the grain, and the second. In most scientific measurements the fundamental units used are now the centimetre, gramme, second. It is therefore of some importance to show how the results in one system are to be expressed in the other.

As a preliminary a few simple illustrations of the theory of physical units will be given, in order to make the subject clear. For additional information the reader is referred to Everett's "Units and Physical Constants." If any length is measured in feet the length may be said to be l' times the length of one foot.

3*

If L' represent the length of one foot, the whole distance may be written $l'L'$. If the same distance be measured in centimetres, l being the number of centimetres and L the length of one, then this distance may also be written lL. As the distance is the same, whatever the system of units used, it follows that

$$lL = l'L';$$

and hence

$$l = \frac{L'}{L} l'. \qquad (47)$$

Here l and l' are the numerical quantities, which are usually called the "lengths," the one being in centimetres, the other in feet. The ratio $\dfrac{L'}{L}$ is the length of a foot in centimetres, or the ratio of the dimensions of these two units. By direct comparison this ratio has been found to be 30.4797. Any length expressed in feet is converted into the equivalent expression of the same length in centimetres by multiplying by this number.

Density is defined to be the mass per unit of volume of any body. It is expressed by the ratio $\dfrac{m}{v}$, where v is the volume of the mass m. If M represents the magnitude of the unit mass, and L the magnitude of the unit length, the magnitude of the unit volume being then L^3, then any density would be completely represented by the expression

$$D = \frac{m}{v} \cdot \frac{M}{L^3}.$$

If M represents the magnitude of the gramme, and L that of the centimetre, the unit density would be a gramme per cubic centimetre. If M represents the kilogramme, and L the decimetre, the unit density would be a kilogramme per cubic decimetre. The unit density in these two cases is identical, and is that of water. If M represents the kilogramme, and L the centimetre, the unit density would be a thousand grammes per cubic centimetre, and would be a thousand times that of water. The density of water in kilogrammes per cubic centimetre is 0.001. Hence the expression $\frac{M}{L^3}$, represents the magnitude or " dimensions " of the unit of density, in precisely the same sense that M represents the magnitude of the unit mass, and L that of the unit length.

If, then, D is the density in grammes per cubic centimetre, and D' the same density in pounds per cubic foot, we shall have an equation similar to the one leading to (47), viz.:

$$D\,\frac{M}{L^3} = D'\,\frac{M'}{L'^3}$$

or

$$D = \frac{M'}{M} \cdot \left(\frac{L}{L'}\right)^3 D'. \qquad (48)$$

If, for example, D be taken as unity, which is that of water (grammes per cubic centimetre), the equivalent density in pounds per cubic foot will be

$$D' = \frac{M}{M'} \cdot \left(\frac{L'}{L}\right)^3.$$

Here

$$\frac{M}{M'} = \frac{\text{gramme}}{\text{pound}} = 0.0022046.$$

Reducing, the value of D' is found to be 62.425, which is the density of water in pounds per cubic foot.

Recurring now to the equations for determining H, viz.:

$$MH = \frac{\pi^2 I}{t^2}$$

$$\frac{M}{H} = \tfrac{1}{2}r^3 \tan u \left(1 - \frac{P}{r^2}\right),$$

it is required to find the conversion factor, which will reduce the value H measured in foot-grain-second units to C. G. S. units. It is to be observed that the value of P is so determined as to satisfy the two equations

$$\frac{M}{H} = \tfrac{1}{2}r^3 \tan u \left(1 - \frac{P}{r^2}\right)$$

$$\frac{M}{H} = \tfrac{1}{2}r'^3 \tan u' \left(1 - \frac{P}{r'^2}\right).$$

Solving these equations for P, its value is found to be

$$P = \frac{r^3 \tan u - r'^3 \tan u'}{r \tan u - r' \tan u'}.$$

Since the tangents in this expression are independent of any system of units, its value can only be changed by a change in the unit of length. Hence the dimensions of P are L^2. The ratio $\frac{P}{r^2}$ will therefore be independent of the system of units used. It will easily be seen that the same is true of the torsion and temperature corrections which have been already discussed.

Solving the two general equations for H,

$$H = A \sqrt{\frac{I}{r^3 t^2}},$$

where

$$A = \sqrt{\frac{2\pi^2}{\tan u \left(1 - \frac{P}{r^2}\right)}}.$$

The value of A is independent of any change in the system of units. I is measured by a mass into a distance squared. r^3 is the cube of a distance. The unit of time is to be the second in both cases. Hence if H' be the horizontal component of the strength of the earth's field measured in the English units, and H the same strength in metric units C. G. S.,

$$H = \left(\frac{M'}{M} \cdot \frac{L}{L'}\right)^{\frac{1}{2}} H'.$$

Substituting the values of the unit ratios,

$$H = 0.046108 \, H'$$
$$H' = 21.688 \, H.$$

$$\log H = 8.863778 + \log H'$$
$$\log H' = 1.336222 + \log H.$$

EXPLANATION OF THE PLATES.

Plate I. represents the U. S. Coast Survey magnetometer, having the deflection-magnet in position for observation. The shorter deflection-magnet is suspended to the torsion-head. The instrument can be used either as a sine or a tangent instrument. It can also be used as a declinometer. In place of the observing telescope of I. a small alt-azimuth instrument like that shown in II. may be used, being mounted on a table-tripod with the magnetometer.

Plate III. represents the form of dip-circle now commonly used. The vertical circle is about six inches in diameter, reading by opposing verniers to minutes. Two simple microscopes serve to read the verniers, while two compound microscopes moving with the verniers are pointed upon the marked ends of the needle. These plates were kindly furnished by the eminent instrument-makers, Fauth & Co., of Washington, D. C., being reproduced from their catalogue.

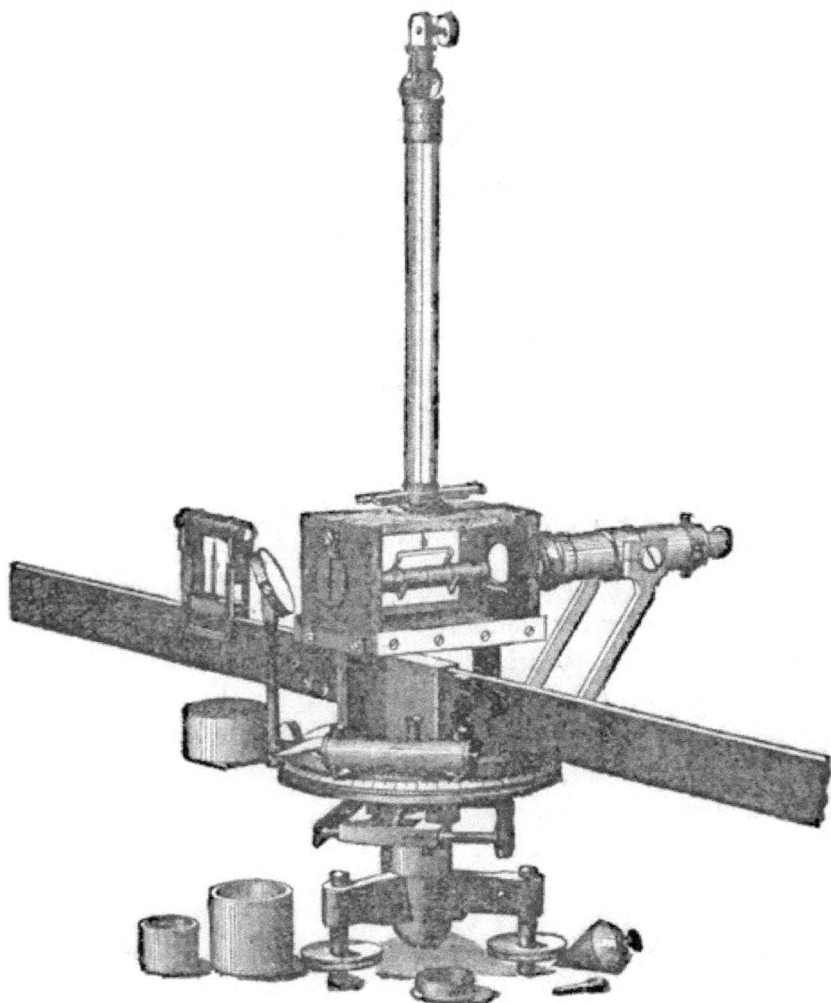

PLATE I

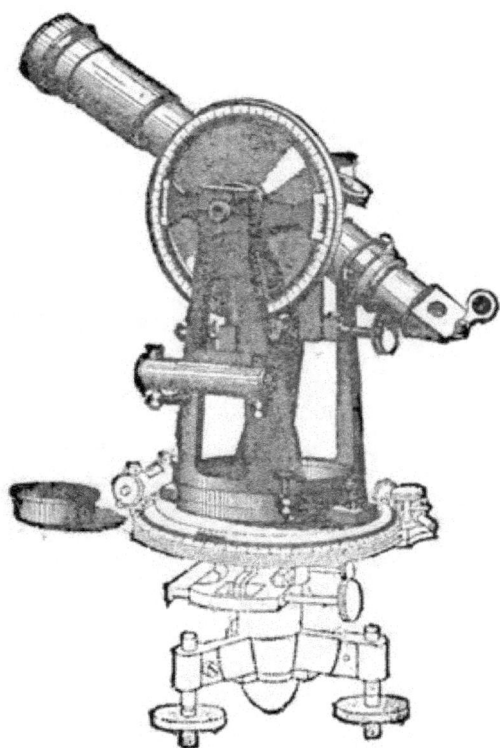

PLATE II.

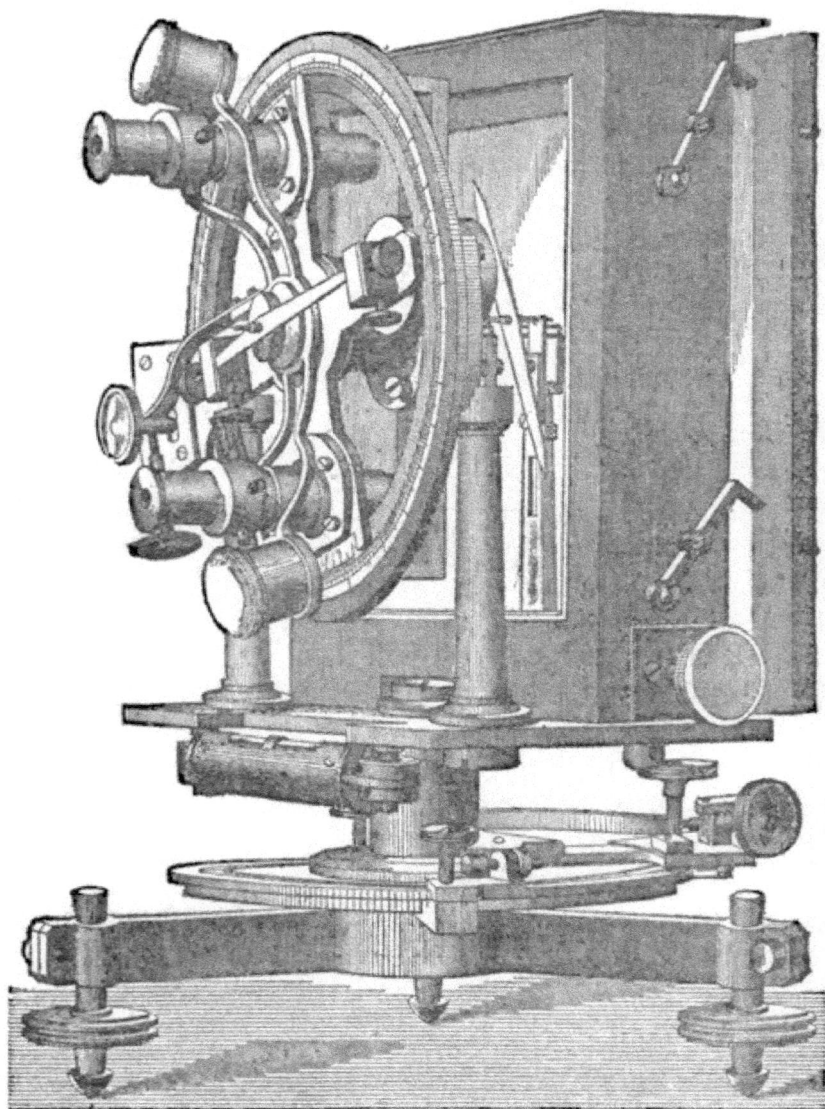

PLATE III.

APPENDIX

ON THE REDUCTION OF OBSERVATIONS BY THE METHOD OF LEAST SQUARES.

If a blacksmith were to repeatedly measure the length of a piece of iron by means of his two-foot rule, his measurements would all agree. If the distance between two fine lines on the bar of iron is measured with the highest attainable accuracy by means of a measuring engine, the results of separate and independent measurements do not agree except by accident, and the tendency to disagree is found to increase with the delicacy of the determination. It is therefore manifest that the true length of the bar can never be obtained. It is conceded universally that the arithmetical mean of the observed values is the best result that can be obtained, when all known corrections have been made and the observations are all equally worthy of confidence, or, in other words, when they have equal weight. Really the observations might be weighted differently by different observers, if one happened to notice something affecting the value of an observation which should escape the attention of the other. Unsuspected causes may thus affect consecutive observations in a different degree, so that observations to which one observer might give equal weight might be differently weighted by another. It is impossible to reproduce exactly the same condi-

tions, and to make consecutive observations in the same
manner. In fact, the differences in the results obtained
are due to such causes. When the observer does not
know of some specific reason for attaching less importance to an observation than to others, it should be
given equal weight, even if it is discordant.

PROPERTIES OF THE ARITHMETICAL MEAN.

Let a single quantity be measured n times, the
observation values being x_1, x_2, x_3, . . . x_n. Let
$x_1 + x_2 + x_3 + \ldots x_n = \Sigma x$. If x_0 represent the arithmetical mean, then

$$\frac{\Sigma x}{n} = x_0. \tag{1}$$

If x_0 be subtracted from each of the observation
values, the resulting differences or residuals will be
sometimes positive, sometimes negative. The smaller
the residuals are numerically, the greater the precision
of the measurements, and the greater the degree of confidence to which the mean is entitled, provided all constant errors affecting all measurements alike have been
previously eliminated. Indicating these residuals by r,
the n observations would give residuals as follows :

$$x_1 - x_0 = r_1$$
$$x_2 - x_0 = r_2$$
$$x_3 - x_0 = r_3$$
$$\text{etc., etc., etc.,}$$
$$x_n - x_0 = r_n$$

By adding, $\quad \Sigma x - n x_0 = \Sigma r \tag{2}$

By (1) it follows that the first member of (2) is equal
to zero, hence

$$\Sigma r = 0. \tag{3}$$

The arithmetical mean renders the sum of the residuals zero.   Any other number thus treated would give
residuals the sum of which would be greater or less
than zero.   If the separate observations were all precisely alike, each residual would be zero.

If the individual equations for the residuals are
squared, the resulting values of r^2 are all positive.   A
little consideration will enable one to see that the sum
of the squares of the residuals obtained from the arithmetical mean will be less than when the residuals are
formed with any other number.   This is, in fact, easily
shown.   The squared equations are :

$$r_1^2 = x_1^2 - 2x_0 x_1 + x_0^2$$
$$r_2^2 = x_2^2 - 2x_0 x_2 + x_0^2$$
$$\text{etc.,} \qquad \text{etc.,} \qquad \text{etc.,}$$
$$r_n^2 = x_n^2 - 2x_0 x_n + x_0^2$$

Adding these equations, the result is

$$\Sigma r^2 = \Sigma x^2 - 2x_0 \Sigma x + n x_0^2.$$

If any other value K were taken instead of the arithmetical mean, the residuals would have different values.
Let them be called ρ.   The last equation would then
read

$$\Sigma \rho^2 = \Sigma x^2 - 2K \Sigma x + nK^2.$$

The value of K is to be found, which will render
$\Sigma \rho^2$ a minimum.   This value is found from the condition

$$\frac{d\Sigma \rho^2}{dK} = -2\Sigma x + 2nK = 0,$$

from which

$$K = \frac{\Sigma x}{n} = x_0.$$

If the different observations are not deserving of equal weight, the reduction must be so made that the values deserving of most confidence shall have most to do with determining the result.

If the variation of the magnetic needle were determined by four measurements to be

9° 24.1 with weight 4
9 25.0   "   "   5
9 24.8   "   "   5
9 22.6   "   "   4

the best value would be obtained by adding in the first and fourth values each four times, the second and third each five times, dividing the sum by the sum of the weights, or 18.

In such a case the weights of the observations might be determined by the judgment of the observer, who is able to cite some specific reason for attaching less importance to some observations than to others. That the result of an observation is discordant is not of itself an important reason, and should be accepted with very great caution, although discordant observations will really affect the weight of the resulting mean.

If the numbers to be weighted are the means of several equally good observations, their weights would in each case be represented by the number of observations. The weight of such a mean is also determined by its probable error, as is shown in more extended treatises, in which the theory of probability is applied to errors of observation.

OBSERVATIONS ON TWO OR MORE QUANTITIES.

Let it be assumed that observations have been directly made on three variable quantities u, v, y, which some graphical or other method has shown to be related, as is indicated in the following equation :

$$y = au + bv. \tag{4}$$

Considering this as representing a physical relation, a and b are constants, the true values of which cannot be determined.   The value of v might be u^2, so that the second member of the equation would represent the first two terms of a series.   It is required to assign values to a and b which will most nearly agree with the observations made on y, u, and v.

Each set of simultaneous observations will give an equation of the form

$$y = u.a + v.b, \tag{5}$$

where y, u, and v now become numerical quantities or coefficients.   Such equations are called "observation equations" or "equations of condition."   The latter term is in more general use, but the former term seems preferable.

Any two of the observation equations would determine the values of a and b.   But, on account of the unavoidable errors of observation, some other set of two would give somewhat different values of a and b.   The values obtained from any two would necessarily satisfy those equations, but they would not in general satisfy the other equations.   In other words, if for each observation equation the value $y - au - bv$ is calculated, assigning to a and b any values determined as above, the value would not in general be zero.   As in the similar

case where the arithmetical mean was treated, each
equation would take the form

$$r_1 = y_1 - u_1\, a - v_1\, b$$
$$r_2 = y_2 - u_2\, a - v_2\, b$$
$$\text{etc.,} \quad \text{etc.,} \quad \text{etc.,}$$
$$r_n = y_n - u_n\, a - v_n\, b$$

where the residuals would not in general be zero for
any values of a and b that could be used. Evidently
any values whatever might be assigned to a and b, and
a set of residuals would result. Values might be chosen so that the residuals might all be positive or they

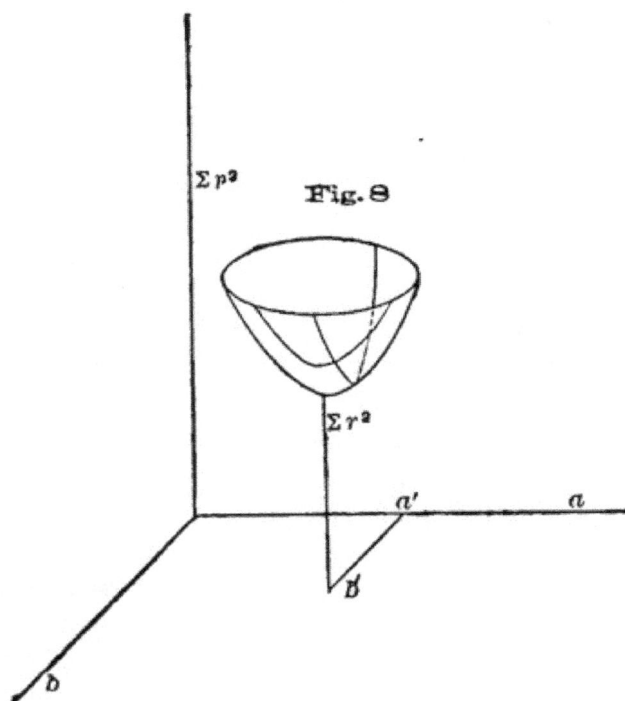

Fig. 8

might all be negative. It is evident that the values of
a and b which will *most nearly* satisfy all the equations

will give small residuals, some of which will be positive
and some negative. Without farther discussion it will
be assumed that the best values of a and b will make the
sum of the squares of the residuals a minimum. The
problem is then to assign values to a and b which will
make Σr^2 a minimum. The values a and b are then to
be treated as independent variables, and with the variable Σv^2 they determine a surface. The conditions for
the minimum point on this surface are:

$$\frac{d\,\Sigma r^2}{da} = 0$$

$$\frac{d\,\Sigma r^2}{db} = 0.$$

The first equation is the condition for the minimum
point in any section of the surface parallel to the plane
determined by the axes of Σr^2 and b. The second is the
condition for minimum on any section at right angles
thereto. If the two conditions are simultaneously im-
posed it determines a minimum point in the surface.
To find the value of Σr^2, the residuals are squared,
which gives the equations :

$r_1^2 =$
$$y_1^2 - 2u_1y_1.a - 2v_1y_1.b + u_1^2.a^2 + 2u_1v_1.ab + v_1^2.b^2$$
$r_2^2 =$
$$y_2^2 - 2u_2y_2.a - 2v_2y_2.b + u_2^2.a^2 + 2u_2v_2.ab + v_2^2.b^2$$
$$\text{etc.}$$
$r_n^2 =$
$$y_n^2 - 2u_ny_n.a - 2v_ny_n.b + u_n^2.a^2 + 2u_nv_n.ab + v_n^2.b^2$$

Adding these equations, the result is

$$\Sigma r^2 = \Sigma y^2 - 2a\,\Sigma uy - 2b\,\Sigma vy$$
$$+ a^2\Sigma u^2 + 2ab\,\Sigma uv + b^2\Sigma v^2. \qquad (6)$$

This is the equation of the surface. The intersection of this surface by the plane of the axes Σv^2 and b is determined by introducing into it the condition $a = 0$. Its equation is therefore

$$\Sigma r^2 = \Sigma y^2 - 2b\,\Sigma\,vy + b^2\,\Sigma v^2,$$

which is the equation of a conic section. Making $b = 0$ in this equation, the value of the intercept on the axis Σr^2 is Σy^2. The minimum point *on this section* is given by

$$\frac{d\,\Sigma r^2}{db} = -2\,\Sigma\,vy + 2b\,\Sigma v^2 = 0,$$

from which

$$b = \frac{\Sigma\,vy}{\Sigma v^2}.$$

The sign of b will here depend upon the sign of the numerator.

It is not necessary to further discuss this surface, as it is only intended to point out the nature of the relation with which we are dealing.

The conditions for the minimum point on the surface are obtained from (6). They are:

$$\left.\begin{array}{l} \dfrac{d\,\Sigma r^2}{da} = a\,\Sigma u^2 + b\,\Sigma\,uv - \Sigma\,uy = 0 \\[3mm] \dfrac{d\,\Sigma r^2}{db} = b\,\Sigma v^2 + a\,\Sigma\,uv - \Sigma\,vy = 0 \end{array}\right\} \qquad (7)$$

The first equation determines a minimum point on any section at right angles to the b axis. The second determines a minimum point on any section at right angles to the a axis. If the two equations are combined by elimination, the values of a and b may be obtained, which determine the minimum point on the surface.

These values are :

$$a = \frac{\Sigma v^2\, \Sigma uy - \Sigma uv\, \Sigma vy}{\Sigma u^2\, \Sigma v^2 - \Sigma uv\, \Sigma uv}$$
$$b = \frac{\Sigma u^2\, \Sigma vy - \Sigma uv\, \Sigma uy}{\Sigma u^2\, \Sigma v^2 - \Sigma uv\, \Sigma uv.}$$

(8)

The calculations are made as is shown in the sub-joined table, in which the observed quantities are given in the first three columns, the quantities called for by equations (8) being the sums of the succeeding columns.

No.	y	u	v	v^2	u^2	vy	vy	uv
1	....	....	....	....	....	....	....	....
2	....	....	....	....	....	....	....	....
3	....	....	....	....	....	....	....	....
.	.	.	.	.	.	.	.	.
.	.	.	.	.	.	.	.	.
.	.	.	.	.	.	.	.	.
n	....	....	....	....	....	....	....	....
	Σy	Σu	Σv	Σv^2	Σu^2	Σuy	Σvy	Σuv

Equations (7) are called "*normal equations*," the first being called the normal equation for a, and the second the normal equation for b. It will be observed that they may be obtained from the n observation equations (5) as follows: To find the normal equation for a, each observation equation is multiplied through by the co-efficient of a in that equation. The sum of the resulting equations is the normal equation for a. Similarly the normal equation for b is obtained by using the co-

4

efficients of b as multipliers, as will be seen by inspecting (5) and (7).

This process is really equivalent to *giving weights to each equation equal to the coefficient of the quantity whose normal equation is to be formed.* These weighted equations are then added together, giving a single equation. If this equation were divided through by n it would give a *properly weighted mean equation.* This division by n is, of course, unnecessary in solving the equations for a and b, as it would not affect their values.

WEIGHTED OBSERVATIONS.

If the observations are not of equal weight they must be additionally weighted. If y_1, u_1, v_1 are each the mean of p_1 equally good observations, y_2, u_2, v_2 the means of p_2 observations, etc., the normal equations become, as will be easily seen,

$$\left.\begin{array}{l} \dfrac{d\,\Sigma r^2}{da} = a\,\Sigma pu^2 + b\,\Sigma puv - \Sigma puy = 0 \\[3mm] \dfrac{d\,\Sigma r^2}{db} = b\,\Sigma pv^2 + a\,\Sigma puv - \Sigma pvy = 0 \end{array}\right\} \quad (9)$$

By elimination the values of a and b are:

$$\left.\begin{array}{l} a = \dfrac{\Sigma pv^2\,\Sigma puy - \Sigma puv\,\Sigma pvy}{\Sigma pu^2\,\Sigma pv^2 - \Sigma puv\,\Sigma puv} \\[4mm] b = \dfrac{\Sigma pu^2\,\Sigma pvy - \Sigma puv\,\Sigma puy}{\Sigma pu^2\,\Sigma pv^2 - \Sigma puv\,\Sigma puv} \end{array}\right\} \quad (10)$$

The computations are made exactly as in the preceding case, excepting that an additional column is added just after the first, containing the weight assigned to each set of values u, v, y. These weights may be assigned wholly on the judgment of the observer, if this is possible, or they may simply be the number of observations of which each is the mean, modified, if deemed proper, as the judgment of the observer may decide.

Where each set of values u, v, y is the mean of a large number of observations, the theory of probabilities enables one to calculate the weights from the probable errors, but ordinarily this is a wholly useless refinement.

If the original function has the form

$$y = a + bv, \qquad (11)$$

the value of u in (4) becomes unity. (In fact, (4) may be put into this form by dividing through by u. The quantity $\frac{y}{u}$ would then replace y, and the quantity $\frac{v}{u}$ would replace v in all subsequent equations, and u would be replaced by unity.)

Putting $u = 1$ or $\Sigma u^2 = n$ in (8), the equations become

$$a = \frac{\Sigma v^2 \Sigma y - \Sigma v \Sigma vy}{n \Sigma v^2 - \Sigma v \Sigma v} \left. \right\}$$
$$b = \frac{n \Sigma vy - \Sigma v \Sigma y}{n \Sigma v^2 - \Sigma v \Sigma v} \qquad (12)$$

The manner of calculating is apparent from the table following equation (8).

If the original function has the form

$$y = bv, \qquad (13)$$

the value of a becomes zero. There can be no normal equation for a, and the first of equations (7) must disappear. In the normal equation for b in (7) the condition $a = 0$ must also be introduced.

The value of b becomes then

$$b = \frac{\Sigma ry}{\Sigma v^2}. \tag{14}$$

It will be observed that this result is different from that obtained by taking the average values of the quantity $\frac{y}{v}$, which would be $\frac{1}{n} \Sigma \frac{y}{v}$. This is simply due to the fact that (14) is the result of weighting the observation equations in proportion to the magnitude of the observed quantities or coefficients y and r in the various equations.

Finally, if observations are made on a single quantity, so that the function has the form

$$y = b, \tag{15}$$

the value of v in (13) and (14) becomes unity and $\Sigma v^2 = n$. Hence (14) becomes

$$b = \frac{\Sigma y}{n}, \tag{16}$$

or the best value of the quantity b is the arithmetical mean of the observations—a result which has already been agreed upon under that head in the early part of the discussion.

The additionally weighted equations corresponding to (12) are

$$\left. \begin{aligned} a &= \frac{\Sigma pv^2 \, \Sigma py - \Sigma pv \, \Sigma pvy}{\Sigma p \, \Sigma pv^2 - \Sigma pv \, \Sigma pv} \\ b &= \frac{\Sigma p \, \Sigma pvy - \Sigma pv \, \Sigma py}{\Sigma p \, \Sigma pv^2 - \Sigma pv \, \Sigma pv} \end{aligned} \right\} \tag{17}$$

For (14) weighted equations give the value of b,

$$b = \frac{\Sigma\, pvy}{\Sigma\, pv^2}, \tag{18}$$

and finally, for weighted observations on a single quantity, (18) becomes

$$b = \frac{\Sigma\, py}{\Sigma\, p}, \tag{19}$$

as was shown in the special example given on p. 76.

GRAPHICAL METHODS.

The preliminary investigation of any new function is always made by graphical methods. In a majority of cases met in practice the graphical method is sufficient. In order to be able to make use of this method, the computer must be familiar with the analytical geometry, so that, from the curve which is obtained by plotting the observed values of the variables, he can form an idea of the mathematical relation sought. By far the greater number of cases which are met in physical investigation are represented by the equation

$$y = bx^n.$$

Here y and x are the physical variables, and b and n are unknown constants, the values of which are to be determined so as to best satisfy the equations. If $n = 0$, then $y = b$, or the function is that given in (15). If $n = 1$, then y is directly proportional to x, and the plotted values of y and x will give a straight line passing through the origin. If $n = -1$, the equation is an inverse proportion, and the curve will be an equilateral hyperbola. If $n = 2$ or $\frac{1}{2}$, the curve will be a parabola, etc.

In some cases it may be necessary to add a constant term to the second member, so that the equation will take the form of (11). To take a special case, in order to make the manner of reduction well understood:

Days.	log M.	y	d
0	9.83989	− 0.00260	+ 134
6	990	− 261	+ 128
20	990	− 261	+ 114
23	9.84055	− 326	+ 111
35	9.83894	− 165	+ 99
50	762	− 33	+ 84
60	868	− 139	+ 74
64	846	− 117	+ 70
74	901	− 172	+ 60
86	766	− 37	+ 48
106	728	+ 1	+ 28
129	778	− 49	+ 5
146	622	+ 107	− 12
156	714	+ 15	− 22
165	784	− 55	− 31
169	700	+ 29	− 35
171	623	+ 106	− 37
184	656	+ 73	− 50
195	679	+ 50	− 61
202	504	+ 225	− 68
218	680	+ 49	− 84
228	568	+ 161	− 94
234	451	+ 278	− 100
245	437	+ 292	− 111
373	242	+ 487	− 239
Means: 134	9.83729		

In 1865–6 Professor Wm. Harkness made a series of intensity determinations, and deduced the log. moment of his magnet at the several temperatures of observa-

tion.* These values were reduced to the mean temperature of his series by means of equation (31). The results are given in the annexed table, where the first
column gives the number of days from his first observation, on Oct. 24, 1865, and the second column the value
of log M at a temperature 75°.8 F.

The values in the two columns being plotted, the
points thus determined are shown on the diagram (p. 89).
It is manifest that if any assumption regarding the decrease in log M be made, it must be that of uniform decrease. The equation representing this relation will be

$$\log M = \log M_0 - ad,$$

where log M_0 is the value of log M at any assumed date,
and d is the number of days from the assumed date to
that of any other observation, a being the daily change
in the value of log M. If the mean of all the values of
log M be taken as log M_0, it gives the value of the quantity log M for the mean date of the series, which is 134
days after the first observation. The straight line representing the observations must run through the point
determined by these two mean values. This line is also
to be so drawn as to give weight to other points in proportion to their distance from the point representing
the mean values. The equations of condition become
of the form

$$\log M_0 - \log M - ad = 0,$$

where log $M_0 = 9.83729$, and where d is estimated in
days from the 134th day, which corresponds to March
17, 1866.

---

* "Smithsonian Contributions to Knowledge," vol. xviii. p. 55
of his memoir.

Calling log M_0 — log $M = y$, the equations of condition become

$$y - ad = 0.$$

These values of y and d are given in the third and fourth columns of the table. In order to form the normal equation for a, each observation equation, of which there are twenty-five, is multiplied through by d, the coefficient of a, and the resulting equations are added. The normal equation becomes

$$\Sigma(yd) - a\Sigma(d^2) = 0.$$

Performing the calculations, the value of a will be found to be

$$a = \frac{\Sigma(yd)}{\Sigma(d^2)} = \frac{3.97497}{203965} = 0.0000195,$$

and hence the original equation becomes

$$\log M = 9.83729 - 0.0000195\, d.$$

It is evident that, after having plotted the values of log M and d, the position of the line can be determined with a precision sufficient for most purposes by means of a fine thread, which is laid through the points in such a way as to agree with them as nearly as possible. The position of this line is shown on the diagram. If desired, one point on the line may be determined with precision by obtaining the means of the observations as in the first two columns of the table. After the line is drawn the slope of the line, or the value of (a), is then found by measuring on the diagram the co-ordinates x', y' and x'', y'' of any two points, which should, of course, be as far apart as possible. In this case

$$a = \frac{y' - y''}{x'' - x'}.$$

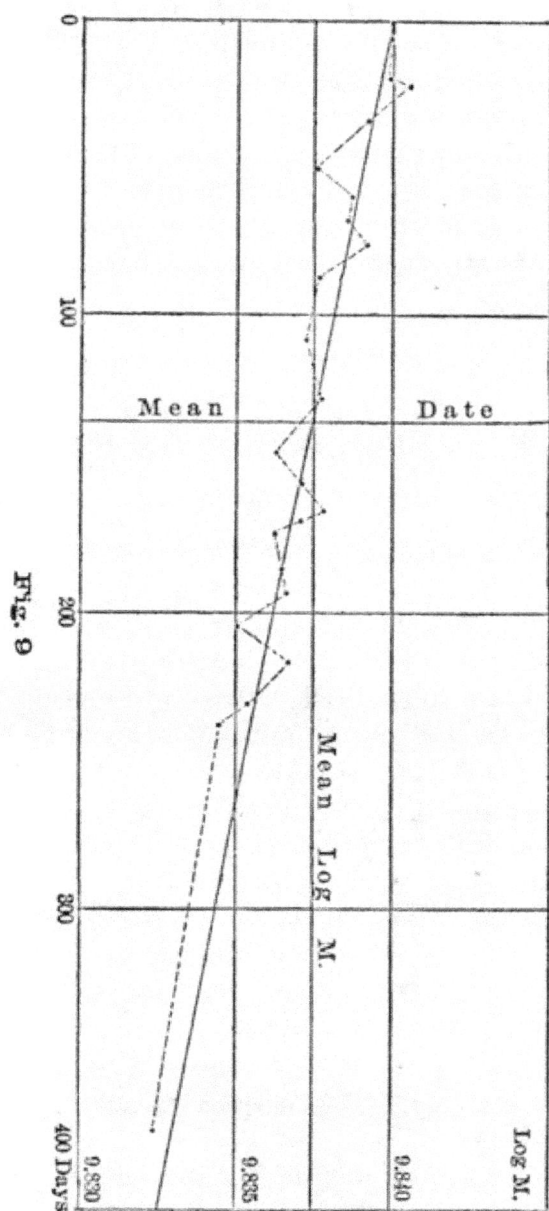

Fig. 9

The co-ordinates are, of course, to be measured in terms of the scales used in plotting.

Such graphical solutions are in the large majority of cases sufficient for all purposes. They should in all cases precede any more exact mathematical solution, in order that one may see whether the observations are sufficiently precise to warrant a more exact solution, or whether the assumed equation is in harmony with the observations.

TIME OF ELONGATION OF POLARIS.

EXPLANATION OF TABLES I., II., AND III.

The following tables (pp. 92–93) give the astronomical times of elongation of the Pole star for the 1st and 15th of each month from 1885 to 1895. They are computed for a latitude of 40° and a longitude of 6^{hrs} from Greenwich. From them the local astronomical time of elongation to the nearest minute, for any latitude between 25° and 55°, may be obtained by applying the correction given in Table II. The correction for difference of longitude is insignificant, amounting to 0^m.15 for a difference of one hour, to be subtracted for places west of the 6th meridian, and added for places east.

To obtain the time of elongation for any date not given in the tables, subtract 3^m.94 from the tabular time of elongation for every day elapsed, if the tabular date is the smaller, or add the same correction if the tabular date is the larger. This correction may be obtained from Table III.

The astronomical day begins at noon, and is twelve hours behind the civil date. From noon to midnight

the astronomical and civil dates are the same ; from
midnight to noon the civil date is one greater.   Thus
Jan. 12, 14^h 40^m, astronomical time is Jan. 13, 2^h 40^m
A.M. civil time.

Example—Required the time of eastern elongation of
Polaris on Aug. 8, 1888, for a place whose latitude is
44° 30′.

Time of elongation, 1888, Aug. 1 (by Table I.).. 10^h 38^m

    Correction for latitude   (by Table II.)..   −1.2

      "   " 7 days   (by Table III.)   −27.6

Time of elongation Aug. 8, 1888............. 10^h 9^m

APPENDIX.

TABLE I.—EASTERN ELONGATIONS.

	1886.	1887.	1888.	1889.	1890.	1891.	1892.	1893.	1894.	1895.
	h m	h m	h m	h m	h m	h m	h m	h m	h m	h m
Apr. 1.	18 38	18 39	18 36	18 38	18 39	18 40	18 37	18 38	18 39	18 40
" 15.	17 43	17 44	17 41	17 43	17 44	17 45	17 42	17 43	17 44	17 45
May 1.	16 40	16 41	16 38	16 40	16 41	16 42	16 39	16 40	16 41	16 43
" 15.	15 45	15 46	15 44	15 45	15 46	15 47	15 44	15 45	15 46	15 48
June 1.	14 38	14 40	14 37	14 38	14 40	14 41	14 38	14 39	14 40	14 41
" 15.	13 44	13 45	13 42	13 43	13 45	13 46	13 43	13 44	13 45	13 46
July 1.	12 41	12 42	12 39	12 41	12 42	12 43	12 40	12 41	12 42	12 44
" 15.	11 46	11 47	11 45	11 46	11 47	11 48	11 45	11 46	11 47	11 48
Aug. 1.	10 40	10 41	10 38	10 39	10 41	10 42	10 39	10 40	10 41	10 42
" 15.	9 45	9 46	9 43	9 44	9 46	9 47	9 44	9 45	9 46	9 47
Sept. 1.	8 38	8 39	8 36	8 38	8 39	8 40	8 37	8 38	8 39	8 41
" 15.	7 43	7 44	7 42	7 43	7 44	7 45	7 42	7 43	7 44	7 46
Oct. 1.	6 40	6 41	6 39	6 40	6 41	6 42	6 39	6 41	6 42	6 43

WESTERN ELONGATIONS.

	1886–7	1887–8	1888–9	1889–90	1890–1	1891–2	1892–3	1893–4	1894–5
	h m	h m	h m	h m	h m	h m	h m	h m	h m
Oct. 1.	18 30	18 31	18 28	18 29	18 31	18 32	18 29	18 30	18 31
" 15.	17 35	17 36	17 33	17 34	17 36	17 37	17 34	17 35	17 36
Nov 1.	16 28	16 29	16 26	16 28	16 29	16 30	16 27	16 28	16 29
" 15.	15 33	15 34	15 31	15 32	15 34	15 35	15 32	15 33	15 34
Dec. 1.	14 30	14 31	14 28	14 29	14 31	14 32	14 29	14 30	14 31
" 15.	13 35	13 36	13 33	13 34	13 36	13 36	13 34	13 35	13 36
Jan. 1.	12 27	12 29	12 26	12 27	12 28	12 29	12 26	12 27	12 29
" 15.	11 32	11 33	11 31	11 32	11 33	11 34	11 31	11 32	11 33
Feb. 1.	10 25	10 26	10 23	10 25	10 26	10 27	10 24	10 25	10 26
" 15.	9 30	9 31	9 28	9 29	9 30	9 32	9 29	9 30	9 31
Mar. 1.	8 35	8 32	8 33	8 34	8 35	8 32	8 33	8 35	8 36
" 15.	7 39	7 36	7 38	7 39	7 40	7 37	7 38	7 39	7 41
Apr. 1.	6 32	6 30	6 31	6 32	6 33	6 30	6 31	6 33	6 34

Table II.	
Latitude.	Correction.
25°	+1ᵐ.9
28	+1 .6
31	+1 .2
4	+0 .8
37	+0 .4
40	+0 .0
43	−0 .5
46	−1 .0
49	−1 .6
52	−2 .3
55	−3 .0

Table III.	
No. days.	Correction.
1	3.9
2	7.9
3	11.8
4	15.8
5	19.7
6	23.6
7	27.6
8	31.5
9	35.5
10	39.4
11	43.3

TABLE IV.—AZIMUTH OF POLARIS AT ELONGATION.

Latitude.	1885.0	1886.0	1887.0	1888.0	1889.0	1890.0	1891.0	1892.0	1893.0	1894.0	1895.0
+35°	1° 26'.4	1° 26'.0	1° 25'.7	1° 25'.3	1° 25'.0	1° 24'.6	1° 24'.3	1° 23'.9	1° 23'.6	1° 23'.2	1° 22'.9
36	27.1	26.7	26.4	26.0	25.7	25.3	25.0	24.6	24.3	23.9	23.6
37	27.8	27.5	27.1	26.8	26.4	26.0	25.7	25.2	25.1	24.7	24.4
38	28.7	28.3	27.9	27.6	27.2	26.8	26.5	26.2	25.8	25.4	25.1
39	29.5	29.0	28.6	28.2	27.9	27.5	27.2	27.0	27.4	27.0	25.8
40	30.3	29.9	29.5	29.2	28.8	28.4	28.1	27.7	29.3	28.9	26.6
41	31.3	30.9	30.5	30.2	29.8	29.4	29.1	28.7	30.3	29.9	27.6
42	32.3	31.9	31.5	31.2	30.8	30.4	30.1	29.7	31.4	31.0	28.9
43	33.3	33.0	32.6	32.3	31.9	31.5	31.1	30.8	32.5	32.3	29.7
44	34.5	34.2	33.8	33.3	32.9	32.6	32.1	31.8	33.6	33.2	30.8
45	35.7	35.4	34.9	34.6	34.0	33.6	33.4	33.0	34.6	34.5	32.0
46	37.0	36.4	36.0	35.6	35.2	35.0	34.6	34.2	36.1	36.1	33.6
47	38.7	37.6	37.2	36.8	36.4	36.0	35.9	35.5	37.5	37.3	34.8
48	40.7	39.3	38.9	38.5	37.7	37.3	36.9	36.6	38.9	38.0	35.7
49	42.7	40.7	41.4	41.0	40.5	38.7	38.3	37.8	40.4	40.5	36.1
+50	45.0	41.8	42.9	42.1	42.0	40.1	41.2	39.3	43.6	41.5	36.7